Hamlyn all-colour paperbacks

Geology

Alec J. Smith PhD FGS

illustrated by
Ralph S. Coventry Associates and
Whitecroft Designs Limited

Hamlyn
London · New York · Sydney · Toronto

FOREWORD

Geology is not one subject but many. It has been described as the application of all sciences in a comprehensive study of the planet on which we live. With the realization of the Earth's limited resources and the growing damage to its environment, some have come to talk of it as 'Spaceship Earth'. It is only natural, then, that more people want to understand their frail home by learning about geology. Unfortunately, many are deterred by the complexity of the subject and the technical jargon of its exponents, and instead they are prepared to regard geology as the collecting of rocks and fossils. This is a pity because the collection of such specimens is really only the very beginning, and these specimens, when understood, have much to tell about events in the history of this planet.

In this book I have attempted to introduce geology, and to show how the many aspects of this large subject are interrelated. It is written for those who know little about geology, and as such, I hope the book will be regarded only as a first step. Size limitations have prevented more than the briefest introduction to minerals, rocks, and fossils. I have concentrated more on the processes, both at the surface and within the Earth; processes which allow this planet, unlike its satellite, the Moon, to continue to show a youthful face.   AJS

Published by The Hamlyn Publishing Group Limited
London · New York · Sydney · Toronto
Astronaut House, Feltham, Middlesex, England

Copyright © The Hamlyn Publishing Group Limited 1974
ISBN 0 600 33932 7

Phototypeset by Filmtype Services Limited, Scarborough, England
Colour Separations by Colour Workshop, Hertford
Printed in Spain by Mateu Cromo, Madrid

CONTENTS

INTRODUCTION

Geology is the science concerned with the study of the Earth – its origins, the processes which cause its development, the beginnings and the evolution of life. Geology is both a theoretical and a practical subject. Geologists use their science not only to learn about the Earth's history, but also in their search for mineral resources in the crust of our planet. They draw upon all the other sciences in their quests.

Geology developed as a science in the late 1700s from a background of mining and mineralogy on the one hand, and from biology on the other. Many significant discoveries have been made in geology since that time, but no development has caught so much public attention as the recently propounded *theory of plate tectonics*. For the first time there is an all-embracing global theory which interrelates many events. In this theory, plates, consisting of whole or parts of continents and ocean floors move relative to each other. These movements cause the splitting of continents and the opening and closing of oceans. In this way mountains are created, earth-

quakes and volcanoes caused, while animal and plant communities may be divided and assume different evolutionary trends. The mechanisms which cause the movements are still a matter of conjecture and are likely to remain so. This is a distinguishing feature of geological science – careful observation, the application of scientific laws, followed by a theory – but the testing of such a theory is not always easy.

For many years geologists relied mainly on a hammer, a hand-lens and a compass-clinometer as aids to the study of the planet for academic and economic purposes. Now, however, a geologist uses a wide range of highly technical and expensive equipment. In future, more such equipment will be developed, because as world population rises and threats to Man's environment grow, the geologist has an increasingly important role to play. He must search for the raw materials that our world demands – oil, coal, uranium, metal ores, water, and even sand and gravel. He must also provide the background information against which changes in climate and contaminants can be measured and compared.

EARTH IN SPACE

Earth is one of the lesser planets of a relatively minor star, the Sun, which is situated in the outer part of a galaxy of stars. Many millions more galaxies exist in the Universe. It is hard to imagine some of the distances involved; these distances are usually expressed in light-years (one light-year is the distance that light travels in one year). Light travels at nearly 300 000 kilometres per second, which means that each light-year represents about ten million million kilometres. Tokyo is about 0·05 light-seconds from London, and the nearest star in the hundred thousand million stars which make up our galaxy, *α-Centauri*, is four light-years distant.

Our galaxy resembles two tightly spiralling arms which form a flattened disc of dust, gas, and stars about 80 000 light-years across. The nearest galaxy to ours is the galaxy, *Andromeda*, about 2·2 million light-years away. It can just be seen by the naked eye, and the light we see left Andromeda before man had evolved.

The orgin of the Universe is a matter of debate, but there is evidence that all the galaxies are retreating from a common point – an expanding Universe created from a core of neutrons which

The place of the solar system in the known Universe.
Above right One possible origin of the solar system.

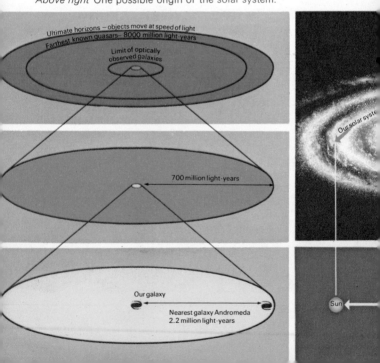

Ultimate horizons – objects move at speed of light

Farthest known quasars – 8000 million light-years

Limit of optically observed galaxies

700 million light-years

Our galaxy

Nearest galaxy Andromeda
2.2 million light-years

Our solar syste

Sun

exploded some ten to fifteen thousand million years ago. Today, we can discern radiation from sources 8000 million light-years away, radiation which started its journey to our radio-telescopes before the Earth existed.

The composition of the Universe is closely related to the matter which makes up our planet. The primary, chemically indivisible substance of matter is the *element*. Ninety-one of all known elements occur naturally on Earth, and over seventy of these have been identified spectroscopically – that is, by the study of light waves – in the Sun. Scientists believe that all elements are constructed from hydrogen, the simplest but most abundant element of the Universe, by thermonuclear reactions in the intensely hot, high pressure interiors of stars. During this reaction energy is produced, much of it as light. Stars evolve by gravitational contraction and thermonuclear burning, their colour changing with age. Some stars in old age explode, creating novae and supernovae which may become the basis of new stars. In this way, about 6000 million years or more ago, our solar system was born.

The solar system consists of the Sun, nine planets, a belt of asteroids, meteorites, and several comets; the main part is about

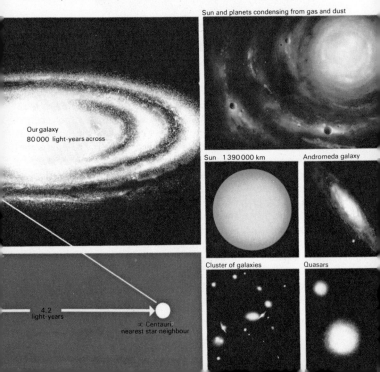

Sun and planets condensing from gas and dust

Our galaxy
80 000 light-years across

Sun 1 390 000 km

Andromeda galaxy

Cluster of galaxies

Quasars

4,2
light-years

∝ Centauri,
nearest star neighbour

SOLAR
FLARE
400 000 km out from sun

The solar system. Key: 1 Mercury; 2 Venus; 3 Earth; 4 Mars;
5 Asteroids; 6 Jupiter; 7 Saturn; 8 Uranus; 9 Neptune; 10 Pluto.

eleven light-hours across. A million times larger than Earth, the
Sun makes up 99 per cent of the mass of the system. It is a vast
nuclear furnace converting four million tons of matter into energy
every second. Great solar prominences flare out from its gaseous
surface which is at a temperature of over 6000 °C. The Sun is the
source of nearly all the energy on Earth. Each hour, Earth receives
solar energy equivalent to burning 20 000 million tons of coal.

Orbiting around the Sun in the same direction as the Sun's
rotation are the planets. The inner planets, Mercury, Venus, Earth,
and Mars, are dense, small planets, while beyond, at ever increasing
distances from the Sun, are the lighter, usually larger planets,
Jupiter, Saturn, Uranus, Neptune, and Pluto.

Earth's planetary neighbours are Mars and Venus. Mars has a
thin atmosphere of carbon dioxide with only the smallest amounts
of water vapour and free oxygen. It has a scarred surface, very
similar to that of our Moon. Venus is permanently covered by

8

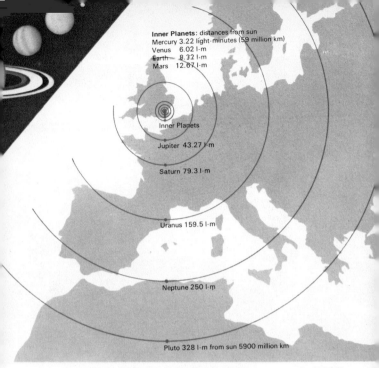

Inner Planets: distances from sun
Mercury 3.22 light-minutes (59 million km)
Venus 6.02 l-m
Earth 8.32 l-m
Mars 12.87 l-m

Inner Planets

Jupiter 43.27 l-m

Saturn 79.3 l-m

Uranus 159.5 l-m

Neptune 250 l-m

Pluto 328 l-m from sun 5900 million km

clouds of carbon dioxide. At its surface temperatures reach 400 °C with pressures four hundred times greater than those at the Earth's surface – conditions which would sterilize living matter.

Earth is fortunate, its rate of rotation makes day and night short which balances temperatures, and its density and position with relation to the Sun enable it to hold an atmosphere. The atmosphere further moderates temperatures and filters harmful rays from the Sun while exerting a pressure sufficient to maintain liquid water at the Earth's surface. The atmosphere and the hydrosphere are the twin harbingers of life. Life as we know it is unlikely to exist on the other planets because the outer planets have too much hydrogen in their atmospheres, and the inner planets are too hot.

Natural satellites revolve around several of the planets; Jupiter has twelve whereas Earth has only one, the Moon. The Moon is about a quarter of the diameter and only $\frac{1}{81}$ the weight of Earth. Its surface is scarred with craters, probably caused by the impact of space debris. The Moon is without atmosphere, it is devoid of life, and therefore it differs markedly from Earth. The Moon is static and

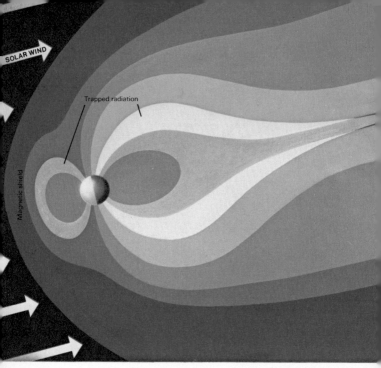

Earth's magnetic shield, an explanation of Earth's magnetic field and the evolution of the atmosphere: A Primordial gases; B Gases from molten surface; C Gases from volcanoes; D Oxygen from plants.

ancient, lacking Earth's secret of eternal youth.

The most widely accepted theory for the origin of the solar system reasons that it was created about 6000 million years ago, through the collapse of a primordial revolving cloud of solid particles, neutral gas, and plasma. In the beginning it was cold, but contraction under gravity made the centre of the Sun hot enough for thermonuclear reactions to start and the planets were heated enough to melt their primary constituents.

At first, Earth must have been much larger. Because of its proximity to the still young Sun, Earth gradually lost its more volatile elements leaving an iron-silicate *magma*. As this cooled creating a crust of *igneous* rock, probably basaltic (see p20) in composition, an early atmosphere developed. Later, less dense granitic (see p20) rocks were also formed and rested on the heavier basaltic rock causing variations in altitude. Water vapour was a common

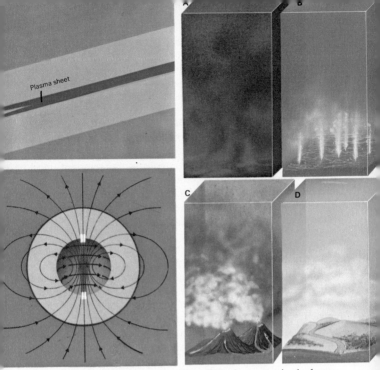

Plasma sheet

product of the cooling magmas and it collected as water in the lower areas to create primitive oceans. Violent rainstorms swept the planet eroding the infant crust creating *sedimentary* rocks (see p66). High temperatures and pressures frequently modified large regions of solidified magma and sedimentary rock to form *metamorphic* rocks. The oldest rocks so far discovered are approximately 3800 million years old; most rocks on Earth's surface are much younger.

The original atmosphere differed from today's: oxygen was not added in quantity until the evolution – some 2000 million years ago – of oxygen-producing plants which used sunlight to break down the carbon dioxide.

The surface of the crust now receives most of its heat from the Sun, but high temperatures still exist within the planet, and Earth has a dense, iron-bearing, molten core. A consequence of this core is Earth's magnetic field, probably the result of a dynamo effect. This magnetism gives Earth protection from charged particles from the Sun – the solar wind. The Van Allen belts are the result of the interaction between Earth's magnetism and the solar wind.

Composition and structure of the Earth

By the fourth century B.C. philosophers had argued that Earth was a sphere, although this was forgotten in the Dark Ages. By the second century B.C. Eratosthenes had determined the size of Earth to within a few per cent of its now known size. Newton, in the 1600s, guessed its density at between five and six times that of water, even though the common rocks at the surface were known to be only about three times that of water. Later work proved the mean density to be 5·52 grammes per cubic centimetre ($5·52 \, g/cm^3$). If the surface is lighter than the mean, then the interior, it was reasoned, must be denser. When Earth was accurately measured, the polar diameter was found to be 22 kilometres (22 km) less than the equatorial diameter of 12 765 km. From this it was deduced that the interior must be part molten with a density as high as $17 \, g/cm^3$ in the inner core.

Most of the detailed evidence concerning the nature of the interior of the Earth has come from the study of earthquakes – *seismology*. Earthquakes have long been studied by man because they are so often catastrophic. The ancient Chinese had a simple system to determine the possible location of distant earthquakes, modern man uses a seismograph. The records, seismograms, frequently contain three distinct sets of shock waves. The first are compressional or primary (P) waves, then come cross vibration or secondary (S) waves, and last oscillatory or longitudinal (L) waves, which often bring destruction. P and S waves travel from the earthquake *focus* in all directions, including through the Earth, while L waves can only travel around the surface of the planet. The speed of travel of all these waves is related to the density of the material through which they are travelling: the higher the density, the greater the speed. Further, S waves cannot travel through a liquid, and their absence in some records is evidence for a molten core. Three seismographs are enough for scientists to locate the *epicentre* of the earthquake which is the point at the surface above the focus. The study of the records made around the world of one earthquake event can tell scientists much about the Earth's interior. Consequently, detailed analysis of the interior has been made by studying the thousands of such records which exist.

The outer layer of the Earth is called the *crust*, and it may be

Above World distribution of earthquake epicentres.
Below Nature and travel paths of earthquake shock waves.

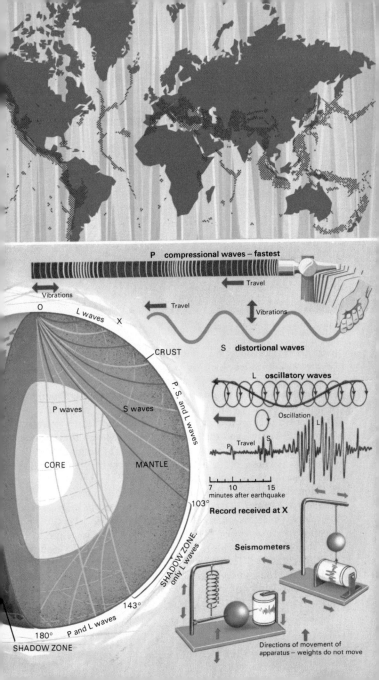

P compressional waves – fastest

Travel

Vibrations

Travel

Vibrations

S distortional waves

Vibrations

O L waves X

CRUST

P. S. and L waves

P waves

S waves

CORE

MANTLE

L oscillatory waves

Oscillation

P Travel S L

7 10 15
minutes after earthquake

Record received at X

Seismometers

103°

SHADOW ZONE.
only L waves

143°

180°

P and L waves

SHADOW ZONE

Directions of movement of
apparatus – weights do not move

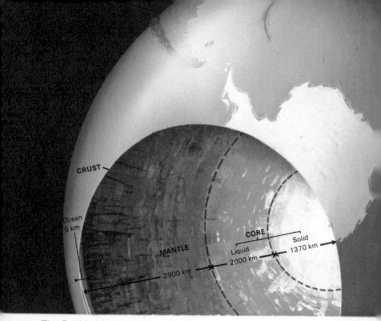

The Earth's interior.

over 50 km thick under the continents. Near the surface there may be several kilometres thickness of sedimentary rocks. These pass downwards into a granite-metamorphic layer, density $2.7 \, \text{g/cm}^3$, often called *sial* after its principal components silica and alumina. Seismic records away from mountains reveal that at about 15 to 20 km down there is a change in the crust. The density increases to about $3.0 \, \text{g/cm}^3$ and the granitic layer gives way to a basaltic layer, called *sima*, after silica and magnesia. The continental crust varies in thickness, but it is always many times thicker than the crust beneath the oceans, where it consists of only about 2 km of sediments and basalt flows, and about 5 km of sima.

Below the sima of continent and ocean lies the *mantle*. The boundary, which, as stated, is much deeper below the continents, is called the *Mohorovicic (Moho or M) discontinuity*, named after the Yugoslavian seismologist who first recognized the evidence for it. There are very few places where mantle rocks can be seen at the surface. The rock, thought to be like peridotite, is heavier than the sima with a density of about $3.4 \, \text{g/cm}^3$. With some variations, the mantle continues for 2900 km to the *core*. The mantle-core boundary can be deduced from the travel paths of P waves. An outer core is believed to be composed mainly of iron and nickel and although it is under intense pressure, the absence of S waves beyond 103 degrees of arc from any earthquake epicentre implies that it is in a liquid state. Densities of 10 to 15 g/cm^3 are estimated for this part of the core. There is also evidence of an inner core starting at 4900 km below the crust which is of similar composition to the outer core, but is solid with densities up to 17 g/cm^3. The iron-nickel composition of the core is, of course, the subject of speculation, but the iron-nickel composition of many meteorites is taken as supporting evidence. The temperature at the core is also the subject of speculation, but 4000 °C is a widely accepted estimate.

Geologists suppose that there are convection currents within the mantle, and that these are the mechanism which moves crustal plates to create continental masses separated by thin oceanic crust, rather than the Earth being a static planet with a crust of uniform composition and thickness.

Contrasting thicknesses of the crust. Key: **1** Granitic layer – sial, density 2·6–2·7 g/cm^3; **2** Basaltic layer – sima, density 3·0 g/cm^3; **3** Mantle – ultrabasic material, density 3·4 g/cm^3.

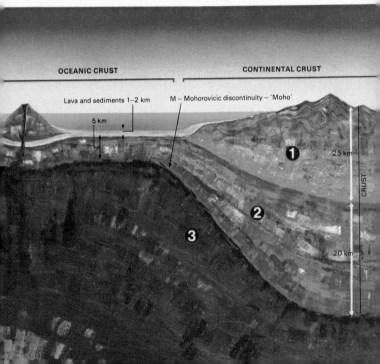

OCEANIC CRUST

CONTINENTAL CRUST

Lava and sediments 1–2 km

M – Mohorovicic discontinuity – 'Moho'

5 km

1

25 km

2

3

20 km

CRUST

ROCKS AND MINERALS

Rocks are the large masses of material which make up the Earth's crust. Some rocks may be glassy but most are made up of discernible minerals. While some rocks may be composed of only one mineral type, (monomineralic), most are made up of a combination of minerals. Minerals are naturally occurring elements or compounds and have characteristic properties (see page 18).

Igneous rocks develop from a *magma* which is molten rock material charged with gases. The principal elements in a magma are silicon and oxygen with, among others, varying amounts of aluminium, iron, sodium, potassium, calcium, and magnesium. On cooling these combine to form mainly silicate minerals and gases. The amount of silica in most magmas ranges between 70 and 40 per cent of the total composition. Magmas high in silica are called *acid*, those low in silica are called *basic*. Not all minerals crystallize at the same time; the illustration shows the order in which the principal silicate minerals crystallize. Petrologists – geologists who study rocks – recognize this as a *reaction series*. As the magma cools, and depending upon its original composition, the high temperature

Reaction series of principal silicate minerals. *Inset* Relation between minerals and classification of igneous rocks (see page 20) Key: 1 Basic; 2 Intermediate; 3 Acid.

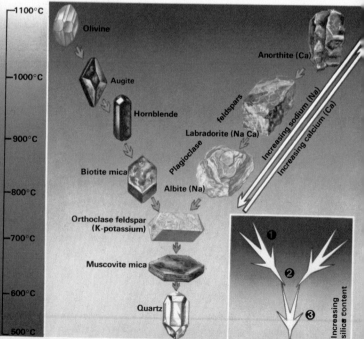

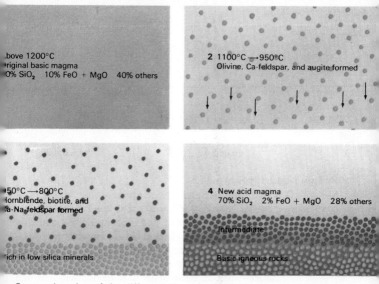

One explanation of the differentiation of magmas by crystal settling.

minerals react with the magma to form new minerals. This sequence of crystallization can also lead to differentiation, when the first minerals to appear sink because they are denser than the remaining magma, allowing a magma to give rise to two or more types of igneous rock.

Sedimentary rocks are derived from the decomposition and disintegration of existing rocks of all types at the surface of the Earth. They can be formed by the settling of particles, by chemical precipitation, and by organic activity. The composition of the original rock, the way the material was carried, as particles or in solution, the method of deposition, and the changes which occur as sediments harden into rock, all exercise a control on the final composition of a sedimentary rock.

Metamorphic rocks are made by the alteration of rocks by heat or pressure or by a combination of both on a regional scale. When temperatures and pressures are particularly high, deep in the crust, rocks may be remelted and in this way become igneous rocks again. Earth is an active planet so that igneous, sedimentary, and metamorphic rocks are constantly being made as part of a continuous cycle of change. This cycle is dealt with on page 28.

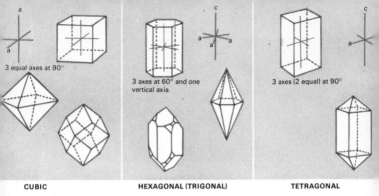

CUBIC — 3 equal axes at 90°

HEXAGONAL (TRIGONAL) — 3 axes at 60° and one vertical axis

TETRAGONAL — 3 axes (2 equal) at 90°

Properties of minerals

Minerals are naturally occurring elements and compounds, and have characteristic chemical, physical, and optical properties. They may consist of a single element, and indeed, different minerals may be composed of the same element. Most minerals are compounds, some of these are simple compounds, such as quartz (silicon and oxygen), others are very complex.

The atomic structure of each mineral is distinctive, and the way the atoms are linked critically controls its properties. Diamond and graphite are both composed of carbon atoms, but diamond is hard because of the strong links between its layers of atoms. The atomic structure also controls a mineral's crystal form, that is the shape it assumes under non-confined conditions. Six crystal systems exist and these are illustrated with some common crystal forms.

Elaborate analyses are necessary to identify many of the 2000 known minerals, but many can be recognized by reference to the following properties:

Hardness – a mineral's resistance to scratching. Ten minerals comprise Mohs' scale of hardness. Talc, the least hard is 1; gypsum

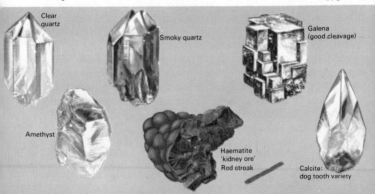

Clear quartz

Smoky quartz

Galena (good cleavage)

Amethyst

Haematite 'kidney ore' Red streak

Calcite; dog tooth variety

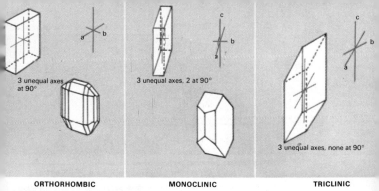

3 unequal axes at 90°	3 unequal axes, 2 at 90°	3 unequal axes, none at 90°
ORTHORHOMBIC	MONOCLINIC	TRICLINIC

Crystal systems showing crystallographic axes and common crystal habits.

2; calcite 3; fluorite 4; apatite 5; orthoclase 6; quartz 7; topaz 8; corundum 9; diamond 10.

Colour – not always a good guide; quartz, for example, has several colours depending upon impurities, and other minerals such as galena (lead-grey) and sulphur (yellow) have constant colours. **Streak** – the colour of the powdered mineral can be more diagnostic, particularly for metallic ores. **Lustre** – the reflected light appearance is also of value: galena is metallic, quartz is vitreous (glassy).

Cleavage – the ability to split on certain planes. A light blow will cleave most minerals: galena has a cubic cleavage, mica a perfect basal cleavage and quartz none. **Fracture** is the way a mineral breaks other than by cleavage: quartz has a conchoidal fracture.

Specific gravity – the relative weight compared with that of an equal volume of water. In minerals, it ranges from 1·7 for borax to 19·3 for gold.

Habit – the common appearance of a mineral: calcite, for example, may occur in several crystal forms, or it may be massive or granular.

Taste and **magnetism** are examples of other properties, but the subject is a big one, and you should refer to the references.

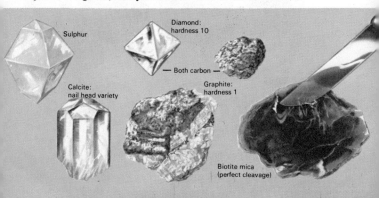

Sulphur

Diamond: hardness 10

Both carbon

Calcite: nail head variety

Graphite: hardness 1

Biotite mica (perfect cleavage)

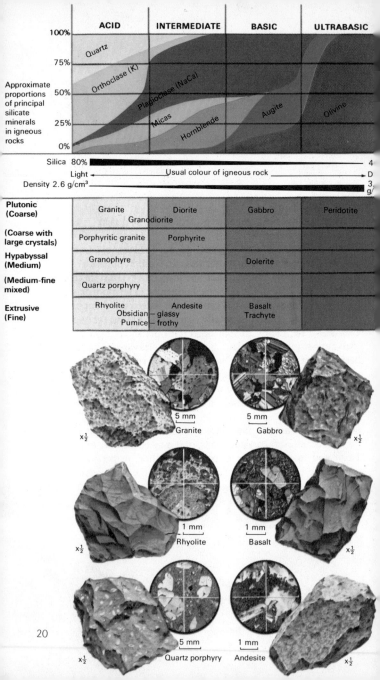

	ACID	INTERMEDIATE	BASIC	ULTRABASIC
Approximate proportions of principal silicate minerals in igneous rocks	Quartz, Orthoclase (K), Plagioclase (NaCa), Micas, Hornblende		Augite	Olivine

100% — 75% — 50% — 25% — 0%

Silica 80% ——————————————————————— 4

Light ← ——— Usual colour of igneous rock ——— → D

Density 2.6 g/cm³ ——————————————————————— 3 g/

	ACID	INTERMEDIATE	BASIC	ULTRABASIC
Plutonic (Coarse)	Granite / Granodiorite	Diorite	Gabbro	Peridotite
(Coarse with large crystals)	Porphyritic granite	Porphyrite		
Hypabyssal (Medium)	Granophyre		Dolerite	
(Medium-fine mixed)	Quartz porphyry			
Extrusive (Fine)	Rhyolite / Obsidian – glassy / Pumice – frothy	Andesite	Basalt / Trachyte	

x½ — 5 mm — Granite

x½ — 5 mm — Gabbro

x½ — 1 mm — Rhyolite

1 mm — Basalt — x½

x½ — 5 mm — Quartz porphyry

1 mm — Andesite — x½

Igneous rocks

Igneous rocks are composed mainly of silicate minerals and are formed from a cooling magma. The simplified classification used here is based on mineral composition and texture. It is a genetic classification because it is based on the origins of the rocks. It can be related readily to the reaction series and differentiation.

The acid rocks, such as granite, have a high silica content, composed mainly of quartz, muscovite, and orthoclase and albite feldspar. The presence of quartz shows that the amount of silica in the cooling magma exceeded that which could be combined to make the other silicate minerals in the reaction series. Igneous rocks containing such 'free' silica are sometimes described as oversaturated. The intermediate rocks, such as diorite, contain biotite, hornblende, and plagioclase feldspars. The basic rocks, such as gabbro, contain mainly augite, olivine, and more calcium rich plagioclase feldspars. Less common, because they have closer affinities to the mantle than to the crust, are the ultrabasic rocks like peridotite: low in silica, they are composed mainly of olivine and anorthite with some augite.

While the minerals reflect the composition and temperature of the original magma, the texture describes the size and shape of the crystals and reflects the cooling history of the rock. Plutonic rocks, such as granite and gabbro, are made up of large crystals because the magma cooled slowly having been intruded at considerable depths below the surface. Such rocks are exposed at the surface only as a result of prolonged erosion. Hypabyssal rocks, like granophyre and dolerite, are composed of smaller crystals, because they are intruded at shallow depths and cool faster. Extrusive rocks cool rapidly on the surface and hence have minute crystals as in basalt, or are, like obsidian, even glassy. Some extrusive rocks, like pumice, have a frothy appearance caused by escaping gases.

The periphery of an intruded magma is often fine textured because it chilled rapidly against the surrounding rocks. Some igneous rocks have a porphyritic texture, large crystals in a fine matrix, caused by the rapid cooling of a partially crystallized magma.

Only a few rock types have been described here, but all the different igneous rocks can be classified by the same methods.

Classification of some common igneous rocks by mineral composition and grain size with some examples in hand-specimen and as seen through a polarizing microscope.

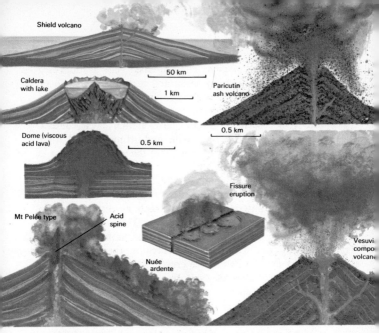

Common shapes of volcanoes.

Extrusive igneous rocks

Molten magmas originate in the mantle or deep in the crust possibly because of radioactive or frictional heating, and because they are less dense than the surrounding rock, they ascend. Magmas are composed largely of silicates and range in composition from acid to very basic. Their viscosity depends upon temperature, composition, and abundance of volatiles, mainly water vapour with lesser amounts of carbon dioxide (CO_2), sulphur dioxide (SO_2), and other gases. If the rise of magma is blocked, gas pressure builds up and the magma can reach the surface with explosive force.

Magma at the Earth's surface is called lava and two types predominate: basalt which is the most abundant, and andesite. The many varieties of lava are derived from these main types by differentiation and contamination. Basaltic lavas, with a low viscosity, can spread over large areas. Old flood or plateau basalts fed from fissures cover large parts of the Deccan of India and the western parts of North America. Similar lavas flowing from a centre produce a low profile shield volcano, such as Mauna Loa in the Hawaiian Islands; standing 9 km above the sea floor, it is over

250 km wide at its base. Iceland is built entirely of shield volcanoes and flood lavas. Andesitic lavas contain more silica, flow less freely, and usually build up a steep composite volcanic cone of lava and pyroclastics – volcanic dust (tuff), bombs, and lapilli (blobs of molten lava). Closely associated are cinder cones, composed entirely of volcanic ash. The circum-Pacific 'ring of fire' is composed of volcanoes of these types.

Basic lavas flow easily, and gas bubbles rise and are preserved as voids or *vesicles*. Thick basaltic flows frequently show columnar jointing, as in the Giant's Causeway (County Antrim, Northern Ireland). Some flows show a surface resembling lines of molten wax, this is called ropy or, in Hawaii, *pahoehoe* lava, while flows with clinker plates on their surface are called *aa* lavas in Hawaii. Lavas extruded below water often show pillow structures caused by the rapid congealing and rupturing of successive masses of lava. Less basic lavas move slowly, often as tumbling masses of irregular blocks, hence the name blocky lavas. Viscous acid lavas are shattered by gases to give brecciated flows. An extreme of this process is the complete tearing apart of the lava causing it to flow as an incandescent cloud of ash and gas, a *nuée ardente*, depositing *ignimbrites*.

Lava flows and fumarolic action.

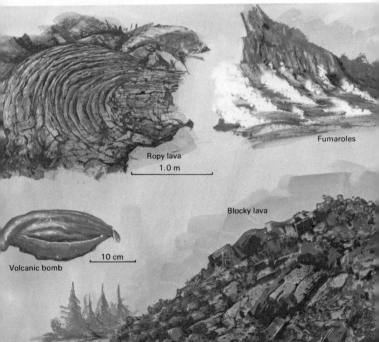

Ropy lava

1.0 m

Fumaroles

Blocky lava

Volcanic bomb

10 cm

World distribution of volcanoes – compare with the map on page 13.

Volcanic land-forms are related to the viscosity of the lavas. Flood basalts cause horizontal surfaces and shield volcanoes have already been described. Composite volcanoes have the classic cone shape of Vesuvius and Fujiyama. Cinder cones, like Paricutin, are steeper. Viscous acid lavas can only push up steep sided domes. An extreme case is when the lava cannot flow but is pushed up as a spine as occurred at Mont Pelée after the 1902 eruption. Many volcanoes have a central crater and some have parasite cones. The collapse of the central part of the volcano can create a wide caldera.

Some volcanoes erupt violently when the volatiles under pressure cause them to explode. Such was the case with Krakatoa in 1883 and Santorini in the fifteenth century B.C.

Fumarolic activity, when only volatiles escape to the surface, characterizes the dormant period of many volcanoes. Water vapour gives rise to juvenile water which is water created for the first time. Sulphur too is deposited near many craters at this time.

Hypabyssal igneous rocks

These are closely related in origin to the extrusive rocks. Intruded at relatively shallow depths, they are now seen at the surface because of the erosion of the cover rocks. Volcanic plugs are roughly cylindrical bodies which represent the necks of volcanoes. Dykes are commonly vertical or steeply inclined bodies of igneous rock, usually of dolerite, which cut across other rocks, that is, they are discordant. They occur in regions of crustal tension. Dykes usually occur together as a swarm, radiating from a centre, roughly parallel or in concentric rings. Sills are of similar composition but are concordant. Laccoliths resemble sills, although the magma moved less freely causing the upper surface to dome up. All intrusive bodies have fine-grained 'chilled' margins, and the adjacent country rocks are metamorphosed.

Hypabyssal intrusions are said to be multiple when showing several episodes of intrusion from one source, and composite when composed of several types of magma. Sills can be confused with old lava flows, because both may exhibit columnar jointing, but the latter often have a vesicular and weathered upper part, while the base of the rocks lying on top show no metamorphism.

Hypabyssal intrusions and contact effects.

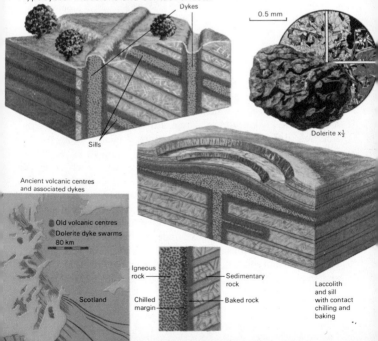

Dykes

0.5 mm

Sills

Dolerite x½

Ancient volcanic centres
and associated dykes

⬤ Old volcanic centres
▨ Dolerite dyke swarms
80 km

Scotland

Igneous
rock

Chilled
margin

Sedimentary
rock

Baked rock

Laccolith
and sill
with contact
chilling and
baking

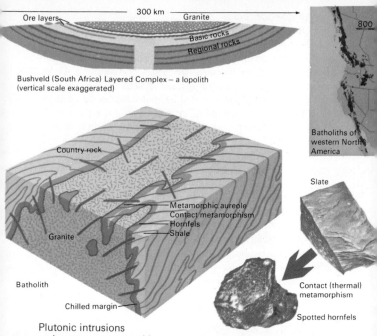

300 km

Ore layers

Granite

Basic rocks

Regional rocks

Bushveld (South Africa) Layered Complex – a lopolith
(vertical scale exaggerated)

800

Batholiths of
western North
America

Country rock

Metamorphic aureole
Contact metamorphism
Hornfels
Shale

Granite

Batholith

Chilled margin

Slate

Contact (thermal)
metamorphism

Spotted hornfels

Plutonic intrusions
and contact metamorphism.

Plutonic igneous rocks

These coarse-grained igneous rocks, formed and intruded at great
depths, cooled only slowly. They are now exposed at the surface
after prolonged erosion. Lopoliths and layered complexes are
saucer-like intrusions which may be several kilometres thick and
as much as 250 km across. These intrusions show signs of considerable differentiation and internal movements. Many of the
lower layers are gabbroic, and layers of minerals of economic
value are often associated with these structures, such as in the
Bushveld complex of South Africa.

Large granitic masses called batholiths may be continuously
exposed for many hundreds of kilometres along the axis of a deeply
eroded mountain chain. They are believed to be created by the
alteration of sedimentary rocks under intense pressures and heat
(regional metamorphism) as the mountain chain was formed.
Shales can be granitized, i.e. altered to slate, phyllite, schist, gneiss,
and finally to granite. Once created, the granite masses can rise,

melting and intruding other rocks. The rocks near the granite are thermally metamorphosed creating an aureole which may be several kilometres wide. In this zone, minerals and rocks are baked and altered so that shales can be transformed into hard hornfels.

Water, rich in minerals and under pressure, is produced in the last stages of granite cooling. It injects the granite, the aureole, and surrounding rock to give veins of tin, copper, lead, and zinc ores.

Hot springs and geysers
These are characteristic of areas of recent igneous activity. Some of the water may be juvenile, but most of it meteoric, and has circulated through the ground and been heated by igneous activity. When the waters reach the surface as hot springs they are rich in minerals which are precipitated as the water cools. Geysers are caused when waters are superheated in underground chambers. Once part of the water changes into steam there is a spectacular display at the surface as steam and water push violently upwards. This intermittent action may occur at regular intervals, as in the case of Old Faithful geyser in the Yellowstone National Park of Wyoming, North America.

Left Geyser *Right* Hot springs and terraces.

GEOLOGICAL CYCLES

Earth's 'secret of eternal youth' was commented upon earlier in this book. The 'secret' depends upon cycles which are recurring series of changes which make Earth resemble a vast machine with many cogs, all working together at different speeds but each one acting upon the others. In this way magmas are brought to the surface, mountains are created and worn away, and rocks decay to form, eventually, new rocks. The last is illustrated in the fundamental rock or geologic cycle; the beginning and the end of this cycle are the rocks at the Earth's surface. Within this cycle are the erosional and hydrological cycles. The former is concerned with the reduction of an uplifted land surface to near sea level. Such a cycle is rarely completed before uplift recurs. Uplift and the variety of erosional agents lead to a diversity of landscapes. The hydrological cycle illustrates the circulation of meteoric water. There are many other cycles – for example, the geochemical cycle illustrates the way in which an element commencing in a magma may circulate through a sequence of geochemical environments.

Through its cycles, the planet is continuously recharging itself,

Landscape showing effects of ice, running water, the sea, and wind. *Inset* The water cycle.

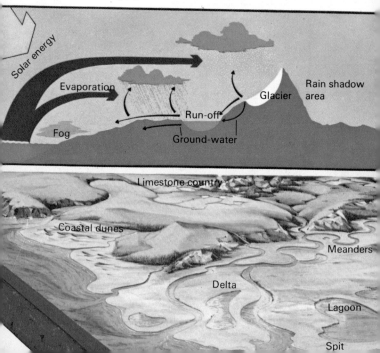

but unfortunately man is using up the Earth's resources faster than they are generated.

The rock cycle

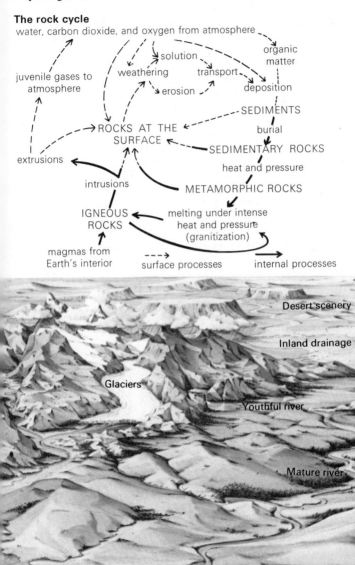

water, carbon dioxide, and oxygen from atmosphere

juvenile gases to atmosphere

solution

weathering

transport

erosion

organic matter

deposition

SEDIMENTS

burial

ROCKS AT THE SURFACE

extrusions

intrusions

SEDIMENTARY ROCKS

heat and pressure

METAMORPHIC ROCKS

IGNEOUS ROCKS

melting under intense heat and pressure (granitization)

magmas from Earth's interior

- - -> surface processes   ──> internal processes

Desert scenery

Inland drainage

Glaciers

Youthful river

Mature river

Tidal flats

SURFACE PROCESSES

All rocks at the Earth's surface are subject to weathering. The products of weathering are eroded and transported as particles or in solution to be later deposited, usually in the sea. Weathering and erosion are denudation processes while the modes of accumulation of sediments are depositional processes.

Weathering occurs at normal surface temperatures. It is brought about by disintegration and decomposition which proceed simultaneously. The former results from mechanical forces – water freezing in small cracks forcing rock apart, rootlets exerting pressures, diurnal heating and cooling weakening a rock by the differential expansion and contraction of its component minerals. Decomposition results from chemical activity. Few minerals are significantly soluble in pure water, but rainwater contains carbon dioxide and is a weak solution of carbonic acid. As such it can dissolve some rocks, particularly limestone. Water containing humic acid from organic matter can also dissolve rocks. Additionally, organic matter draws nutrients from rocks, aiding decompositional changes in many minerals, causing them to expand, and promoting disintegration. The cracks caused by disintegration facilitate the movement of these dissolving waters. Atmospheric pollution hastens weathering processes.

Frost shattering leads to sharp, shattered rocks where the expansion of freezing water prises the rock apart along its joints. In warm climates heating and cooling of a rock may cause exfoliation by rapid expansion and contraction.

The rate of decomposition of the gravestone in the atmosphere is indicated by the legibility of the inscription. Trees and plants can aid in mechanical weathering.

A granite mass can disintegrate into an arkose which is a sandstone of granite composition, or decompose to give quartz sands, muscovite mica, and clays, with some of the original elements going into solution. Basic rocks decompose more readily and more completely than acid rocks. In arctic regions and at high altitudes disintegration is more in evidence than decomposition; in contrast, decomposition predominates in tropical and subtropical regions. When free water exists decomposition is hastened, and soluble salts are carried away by groundwater to rivers. Occasionally, these salts are redeposited near the surface of rocks, hardening them. Where frost action has occurred, sharp, shattered rocks abound. Elsewhere more rounded masses develop as irregularities are weathered away. This may be partly due to exfoliation, or onion-skin weathering, where the sheets of rock split away leaving a spherical form. Weathering reduces rock to a weathered cover, the regolith, preparing rock for erosion and transport.

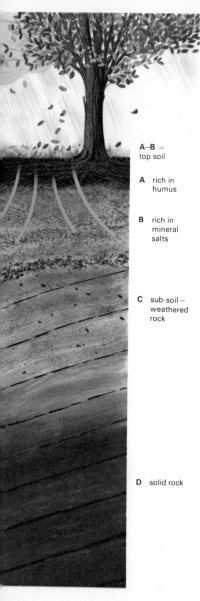

A–B = top soil

A   rich in humus

B   rich in mineral salts

C   sub-soil – weathered rock

D   solid rock

Soils

Most people regard soil as a mixture of humus derived from decayed organic matter, and mineral particles. The latter are derived either directly from weathered rock giving residual soils or from transported debris giving transported soils. Geologists also include the subsoil, the weathered material which rests on the solid rock. Engineers use the term for any weak deposits, and often call soft rocks 'soil'. Pedologists – soil scientists – refer to a soil profile: at the base, solid rock, the D horizon, then subsoil (C horizon), and above soil which is divided into an upper A horizon, often rich in humus but leached of mineral salts by rainwater, and a lower B horizon, which has less humus and where the leached salts are deposited with fine silt and clay particles.

While soils often reflect the composition of the parent rock, pedologists recognize that the formation of particular soils is influenced by climate which controls the type of weathering, the movement of subsurface water, and the vegetation.

Tundra soils of the Arctic have a peaty surface layer on blue-grey, acid, waterlogged clay which rests on frozen subsoil or

Soil profile typical of moist temperate regions. Arrows show direction of moisture movement.

Tundra profile

Peat (A)

Clay (B)

Frozen ground

Terra rossa

Red clay (A)

Lime-stone (D)

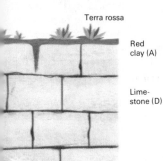

Laterite

Clay (A)

Concretions rich in iron (B)

C

Leached zone rock (D)

Desert soil

Sand (A) (no humus)

Hard pan (B)

C

D

rock. In cool temperate areas grey *podzol* soils result from strong leaching, but some good temperate soils develop, such as the *brown earth* of Europe where a once extensive deciduous forest cover produced good humus, and the *black earth* of the Ukraine and prairies where seasonal rains on grassland produced a rich humus layer.

In warmer climates subsurface moisture can move upward in drier seasons and may produce a *hard pan*, which is a layer of carbonate minerals in the B horizon and is a problem for farmers. A similar feature occurs in some desert areas giving rise to *caliche* layers near the surface. In tropical regions, heavy seasonal rains and intense evaporation can produce *laterite* by the creation of iron oxide concretions – concentric precipitations around a centre – in the B horizon. In special situations, *bauxite* develops with concentrations of aluminium oxide rather than iron oxide.

Localized soils develop in special situations: in warm regions the insoluble remains of limestones accumulate to give thin *terra rossa* soils. Other special soils exist near volcanoes and on wind-blown deposits such as loess.

Varieties of soil profiles.

Denudation

Circulating groundwater is constantly dissolving the regolith, and erosion removes the insoluble debris. The Sun and gravity drive the erosional agents, be they running water, ice, wind, or the sea. Gravity also acts directly to move debris downslope.

Gravity transport ranges from the almost imperceptible to very rapid movements. The slow movement is called creep and is aided by raindrops hitting the soil, by root action, frost heave, swelling by wetting, by burrowing creatures, and disturbance by grazing animals. The rate of movement can be influenced by the type of vegetation cover and by farming practice. Solifluction results from the thawing of frost-heaved soils, the saturated debris spreading out downslope in sheets and tongues. Mudflows are particularly rapid movements of water saturated debris; a special variety is the *lahar* caused by torrential rain on pyroclastic deposits around volcanoes.

Rainwash can rapidly remove soil and soft rock. Exceptionally, scattered boulders can protect patches of ground creating earth pillars. Landslides or slumps develop when masses of weathered material and/or solid rock move downslope faster than general creep. Each landslide possesses distinct boundaries: upslope a gaping scar, cracks or ridges along its sides, and the surface becomes irregular and jumbled as it spreads downslope. The mass moves on a slide

Earth pillars and screes *Inset* Impact of raindrop on wet soil.

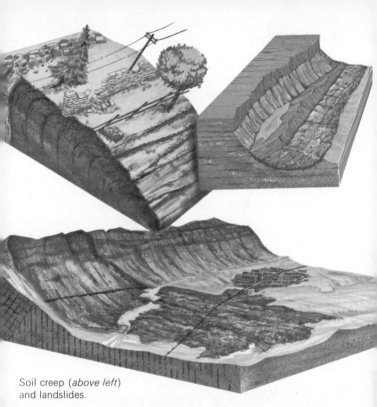

Soil creep (*above left*)
and landslides.

plane. In uniform material such as clays, the whole mass rotates on a curved plane. When movements involve solid rock, the slide usually occurs on weak planes within the rock. Landsliding often follows heavy rains which increase the weight of slopes and reduce friction.

Large rock falls occasionally crash down steep slopes, smaller falls occur where the rocks are subject to disintegration. The coarse angular debris accumulates at the foot of the slope as a scree or *talus*. In high mountains, masses of snow, ice, and rock move downslope as avalanches. These move rapidly on a cushion of trapped air. When they reach lower altitudes the snow and ice melt and, as often happens in the Andes, destructive mudflows result.

Large slumps also occur under water; here the debris mixes with the water causing turbidity currents which, being denser than the clear water, sweep swiftly down submarine slopes.

Running water

When rain falls on to a surface some of the water percolates into the subsoil and rock to become groundwater, some is evaporated back into the atmosphere, and that which remains as surface water becomes run-off. In a rainstorm thousands of tons of water fall on each square kilometre. As the impact of raindrops loosens surface particles, the drops coalesce to make sheets of water carrying particles downslope. This sheet flow does not travel far before developing a channel, often along a ploughed furrow or a track. Deep gullies may develop, creating badlands where there is deep weathering. Water swirls down channels deepening them and undermining banks which slump into the moving water. When the rain stops, the flow soon ceases and the debris is deposited as alluvium. Permanent streams develop when groundwater issuing from springs enables them to flow during dry periods.

Running water is a powerful eroding force; besides dissolving rocks it deepens and widens its channel by frictional drag. It can abrade the channel bed by swirling pebbles to scour potholes. Moving water is turbulent and can carry particles in suspension, by saltation, that is, the bouncing along of larger particles, and by bed load, that is, the rolling and dragging of still larger fragments.

Above left Sheet wash and gullying *Above right* Badlands — advanced erosion by rain *Below left* Ripples formed by scour and deposition *Below right* Pot holes seen in dried river bed.

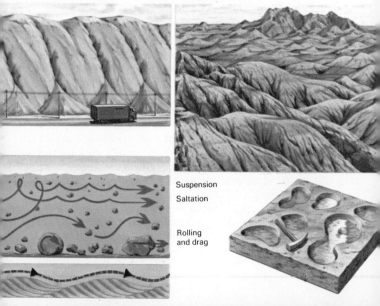

Suspension

Saltation

Rolling and drag

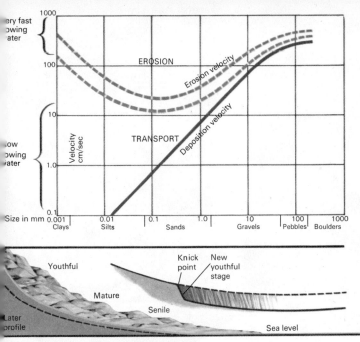

Above Relationship between erosion and deposition of particles in moving water *Below* Idealized profile of river erosion *Inset* Rejuvenation effect of falling base level.

All particles are worn down, sorted, and rounded. The size and quantity of the debris which comprises the load of a stream depend upon its volume and velocity. The velocity depends upon the slope of the stream bed. A balance exists between erosion and deposition: a stream not fully laden will erode, but should the load become too great, due to reduced velocity, for example, then deposition will occur. It is for these reasons that a stream in flood erodes vigorously, is discoloured by its load, and can move large boulders, while at other times the same stream is often crystal clear, its load temporarily deposited.

Streams join together to form a river, the whole making a river system. The longitudinal profile or *thalweg* of such a system is roughly common to many systems and represents a balance between erosion, velocity, volume, particle size, and deposition. The profile is controlled by a base level, usually the sea. River erosion can reduce a landscape to a nearly flat surface known as a peneplain.

Waterfall caused by resistant rock

1

2

River terraces

Flood plain

3

Natural levees

Formation of meanders and oxbow lake

Streams and rivers

Four stages: incipient, youthful, mature, and senile, are recognized in the evolution of a river system. They occur together in a system but given time only the senile stage will remain.

The incipient stage is usually seen in upland areas where short streams with ill-defined courses run between marshy patches. Soon, a distinct valley develops and the stream is in the youthful stage. In cross-section the valley resembles a narrow v. It is rarely straight and interlocking spurs are common. Waterfalls, cascades, and small lakes abound, and there are many tributaries. The water flow is irregular, and large boulders and angular pebbles are scattered in the bed. Rapid deposition can cause braided, dividing, streams and local alluvial cones. The interfluves, the areas between the valleys, are broad and poorly drained.

Downstream the valley widens to a broad v with a flat bottom as the river enters the mature stage. The river no longer occupies the full width of the floor, because a flood plain covered with alluvium has developed and the river cuts successive terraces. The sediments are finer and better sorted. As the river swings on the flood plain it cuts on the outside of bends and deposits on the inside to start meanders. The interfluves are now rounded and the tributaries fewer.

In the senile stage, the river is very wide, its meanders swing across a broad flood plain, and the channel may even be raised above the plain by levees which are natural embankments built up by the river in times of flood. Usually the river flows slowly carrying only fine sediments. The interfluves are low with isolated hills, called monadnocks, left by erosion. Successive swings of meanders can join to cut off a loop creating an oxbow lake. This stage is controlled by a base level, usually the sea, but it may be of hard rock. If it was the latter, erosion downstream would be of the rapid youthful type. Should the base level be removed, as would happen if sea level fell in relation to the land, then each stage would be progressively rejuvenated. At the point of rejuvenation, the 'knick' point, there may be a waterfall or rapids, and a sharp youthful v would be cut into the floor of the old valley. In special circumstances previously developed meanders have been deeply incised following the uplift of a peneplain.

Development of river scenery 1 Youthful river 2 Mature river 3 Senile river.

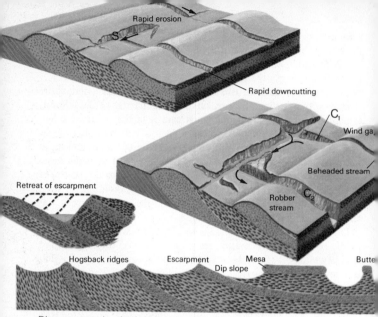

River capture by the headward erosion of the subsequent stream S towards the consequent C_1 eventually capturing its waters so that C_2 becomes more vigorous and down-cutting increases.

River drainage patterns

When a surface emerges from the sea a pattern of drainage is quickly formed. If the rocks which make the new surface are uniform in composition a random *dendritic* pattern, like the branches of a tree, will be established. Usually, however, the rocks will be in bands of differing resistance to erosion, and while the first *consequent* streams, will flow down the original slope, later *subsequent* streams will follow the softer bands and may be at right angles to the consequent streams. In this way a *trellis* pattern is established. This type of erosion leads to the development of steep escarpments and low-angle dip slopes when the strata are of alternating hardness and gently inclined. Streams will tend to undercut the escarpment by moving down the dip slope of the harder rocks giving rise to an asymmetric form. Other features of related origin are the steep sided hogsback, the flat-topped mesa, and smaller butte. In this way rivers are often the cause of outliers, areas of younger rock entirely surrounded by older rocks.

Many rivers erode rapidly along lines of weakness caused by

rock movements. Should such circumstances permit one stream system to cut down faster than another, *river capture* can occur diverting the headwaters of one stream into the 'robber' stream, leaving a beheaded stream as a misfit in its old course.

While most drainage patterns reflect the geological setting, a pattern can be established on a cover of one type and, having worn it away, be superimposed on to a quite different geological setting. The eroding ability of rivers is well illustrated in *antecedent* drainage – here a river was established before a geological event and is not deflected by the event. A fine example is the Brahmaputra River which maintained its channel cutting a deep gorge while the Himalayas were uplifted athwart its course. Both superimposed and antecedent drainage are examples of *inconsequent* patterns.

River courses can be deflected; for example, an episode of glaciation (see later) can create new channels and block old ones, leading to a new pattern of drainage for a region. Steep-sided valleys called gorges are often the result of such new situations caused when valley widening processes could not keep pace with the vigorous downcutting.

Above Superimposed drainage *Below* Antecedent drainage.

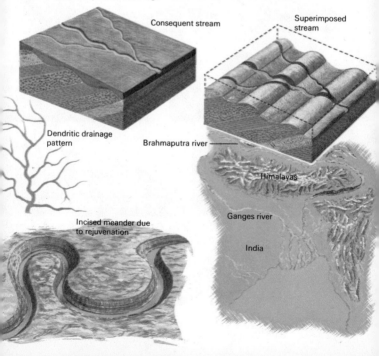

Consequent stream

Superimposed stream

Dendritic drainage pattern

Brahmaputra river

Himalayas

Ganges river

India

Incised meander due to rejuvenation

Underground water

Water percolating into the ground becomes groundwater. Some water is held in the soil, but much of it moves through the subsoil into the rock. Most rocks have voids in them although these decrease with depth. Passing through a zone of aeration, water reaches a zone of saturation where all the spaces are water filled.

Permeable rocks permit the passage of water through pore spaces, and along cracks. Rocks permitting the passage of water but not through pores are pervious. The presence of pores does not necessarily mean that water can pass through these pores. Porosity, the ratio of pore volume to total rock, may reach 50 per cent in clays, but clays are impervious because the pore spaces are extremely small. Clay has no cracks in it and is impermeable. Sandstones have only 15 per cent porosity, but because of the large pore spaces, they are usually pervious. Chalk is very porous and fine grained, has cracks and hollows in it and is permeable although not pervious.

The water-table and examples of the geological control of the occurrence of springs.

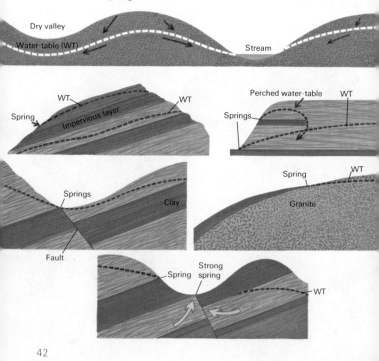

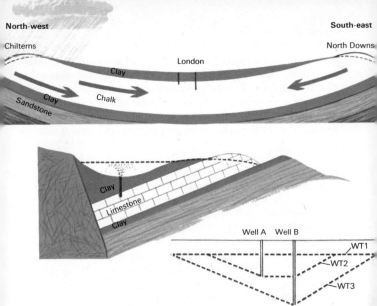

Above Rain falling over the Chilterns percolates through the chalk to provide a water supply for London *Centre* One type of artesian well *Below* Pumping water from Well B lowers the water-table and Well A dries up.

Rocks bearing water make an aquifer. The top of the aquifer is called the water-table. This will rise after rain and fall in dry seasons. It reflects, in a subdued form, the topography, being higher under hills than in valleys. The water in the aquifer flows, emerging where the water-table reaches the surface at rivers, lakes, and springs. An aquifer may rest on an impermeable layer that overlies unsaturated material, creating a perched aquifer and water-table.

Water is recovered from an aquifer by means of a well. Excessive extraction of water can rapidly lower the water-table and in extreme cases dry up adjacent, shallower wells. The replenishment rate depends upon the speed of water flow through an aquifer. This can vary from a few metres a year to several hundred metres a day.

If an aquifer is overlain by an impermeable layer, the aquifer may be confined. If the water-table at the intake to the aquifer is above the top of a well dug through the impermeable layer, then the water will rise creating an artesian well. Large artesian basins exist in many parts of the world and their intakes may be many kilometres from the well.

Caves and limestone topography

Pure water cannot dissolve the calcium carbonate which constitutes limestone, but rain water, being slightly acid and assisted by humic acids from soils, can effectively remove calcium carbonate in the bicarbonate form. It enlarges cracks and fissures to create caves, which are underground systems of irregular passages and chambers. Solution may be so extensive that no surface drainage remains; instead, the water passes underground through sinks or swallow holes.

Caves may reach deep underground; French cavers have explored to depths of more than a kilometre. Some chambers can be of enormous proportions; the Big Room of the Carlsbad Caves in New Mexico, for example, is over a kilometre long, 200 metres wide, and nearly 100 metres high. Such chambers are made by a combination of solution, erosion, and roof collapse. In extreme cases the collapse affects the surface making gorges. Where surface and underground solution and erosion together with collapse are extensive, a distinctive *karst* landscape of dry valleys, elongate closed depressions, sink holes, cave openings and pavements of bare limestone with open joints (grikes) and blocks (clints) is

Limestone topography.

Swallow hole

Water-table

Cavern system

Subterranean lake

Cave painting

formed. The name 'karst' comes from a district in Yugoslavia where this landscape is well developed. Similar areas exist in the Cumberland Plateau in North America, the Causses of southern France, parts of Australia's Blue Mountains, and in the south Pennines and Mendip Hills of England. At the edge of this landscape rivers emerge as large springs, and river erosion is resumed.

Cave deposits are distinctive. Stalactites and stalagmites are formed by the evaporation of dripping water rich in calcium carbonate called hard water. Sedimentary particles are also deposited in caves and these often contain the remains of cave inhabitants, including those of early man.

Chalk, a soft, white variety of limestone, exposed in southern England and northern France does not give rise to karstic landscape. Instead the hills are rounded and grass-covered, and dry valleys are gentle and open. It is possible that these valleys were cut at a time of higher water-table; indeed, extensive surface erosion may have occurred in the closing stages of the Ice Age when underground water was frozen. In some areas, clay with flints, that is, the insoluble residue from chalk, forms an extensive associated deposit.

Rock pavements

Steep gorge

Stalactites

Stalagmites

Ice
Moving ice

At high altitudes and latitudes, precipitation is usually in the form of snow. Above the snow line there is little summer melting; snow is compressed into ice which, under the load of more ice, flows like a viscous fluid carrying the upper fractured or crevassed ice with it. The moving ice is called a glacier and it will continue to flow to a point where the melting rate equals the supply of ice. At present, ice sheets are restricted to Polar regions, including Greenland, and to high altitudes in what are now temperate regions. During the recent Ice Age, continental ice sheets, over a kilometre thick, covered large parts of the northern continents, and valley glaciers radiated from mountain ice fields.

Moving ice is a powerful eroding agent, removing weathered material and plucking and prising fragments from the rock floor. The fragments scrape and smooth the rock and are themselves shattered and abraded. Moving ice modifies the preglacial land-

Left Glaciers flow down-valley from the snow-field and many coalesce into piedmont glaciers.
Right Extent of Quaternary ice sheets shown in white.
Inset The centre of a glacier moves more quickly than the edges.

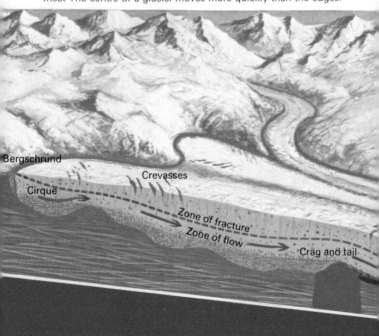

scape and this is particularly well shown in valley glaciers. At the head of the valley, the ice erodes an armchair-like hollow, known as a *cirque*, *cwm*, or *corrie*. This hollow is often overdeepened, because ice can move uphill if pushed by a sufficient mass of ice, and today the hollow is frequently the site of a lake. The back wall is always steep. When ice occupied the hollow there was usually a crack, a *bergschrund*, between the ice and the rock where frost shattering was prevalent. When two cirques back on to each other an *arête* or *striding edge* is formed, while where three or more cirques meet a *pyramidal peak* or *horn* is developed. Main valleys are cut down deeply changing from v to u-shaped cross-section, leaving shoulders of the old valley floor as an *alp*. Former tributaries are left hanging by the rapid down-cutting, and the old interlocking spurs are truncated. *Striated pavements* which are bare rock marked by the passage of ice, and *roches moutonnées* which are rock masses smoothed by ice on the up-valley side and rugged from glacial plucking on the down-valley side, are typical features. Over-deepenings of the main valley due to weaker strata or valley narrowing which caused the glacier to accelerate, become the later sites of lakes. *Fjords* which are deep coastal inlets, usually with shallow seaward rills, are also formed by glacial overdeepening.

Moraine   Melting front   Ice today

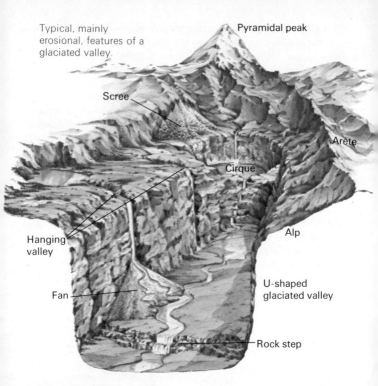

Typical, mainly erosional, features of a glaciated valley.

Pyramidal peak

Scree

Arête

Cirque

Hanging valley

Alp

Fan

U-shaped glaciated valley

Rock step

Glacial deposits

Rock fragments falling from the valley sides on to a glacier are carried along at the edge of the ice as marginal or lateral *moraines*. When two glaciers join, marginal moraines combine into a common medial moraine which makes a ribbon of debris on the surface of the united glacier: a major glacier with numerous tributaries has many medial moraines. Part of the debris falls into the ice and joins that plucked off the rock floor. Some of this load may be deposited below the moving glacier as ground moraine. All the load carried by a glacier is deposited to form a terminal or end moraine at the point where melting balances the arrival of the ice. Ice continues to move forward even when the ice front is retreating. During a retreat recessional moraines, a succession of terminal moraines, may develop together with ground moraine and dumped lateral and medial moraines. These morainic deposits are called *glacial drift* or *till*, or sometimes *boulder clay* in an attempt to describe the nature of the material. The deposits are unsorted ranging in size from boulders to *rock flour*.

Drumlins are revealed by the retreat of the ice; they are low, rounded hills, sometimes more than a kilometre long, aligned and tapered in the direction of ice flow under which they were deposited. They occur in groups and give a 'basket of eggs' topography. Another glacial feature is a *crag and tail*, caused by tail of glacial drift deposited in the lee of a resistant body. Ice transported blocks, alien to the rocks on which they now rest, are called *erratics*. They can often be traced, as a train of blocks, back to their source.

Whether the ice front is stationary or retreating, melt-waters spread *fluvioglacial* sediments to make large outwash plains crossed by braided streams. In dry times these plains are scoured by wind erosion. Near the ice front, melt-water springs build up debris fans called *kames*. Sinuous gravel ridges, known as *eskers*, are deposited from former sub-glacial streams. Blocks of ice left when the glacier retreats make *pingo* hills. In time this ice melts to form rounded depressions known as *kettle holes*. As ice retreats there are many lakes in which *glaciolacustrine* sedimentation occurs, including *varved sediments*. These are made up of seasonally controlled alternating layers of silt and clay, which respectively represent the spring thaw and winter icing over of the lakes.

Typical, mainly depositional, features of a glaciated region.

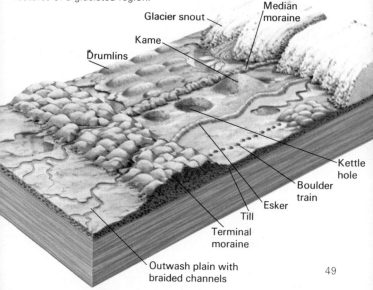

Median moraine

Glacier snout

Kame

Drumlins

Kettle hole

Boulder train

Esker

Till

Terminal moraine

Outwash plain with braided channels

49

The Ice Age

When the ice sheets of the Ice Age finally retreated some 20 000 years ago, they left an extensively modified landscape. A new set of landforms had been created and glacial debris blocked some valleys creating lakes, and caused some rivers to take new courses. As the ice melted it enlarged some rivers which cut deep valleys, carried heavy sediment loads, and formed broad terrace deposits.

There is no general agreement about the cause of the Ice Age which commenced about two million years ago. Four major ice advances are widely recognized with three warm interglacial stages. Indeed, the Ice Age may not yet be over, and we may be living in a fourth interglacial stage! There is no satisfactory explanation for the cause of the low temperatures; some scientists suggest a fall in the solar heat reaching the Earth's surface, and others believe there were changes in ocean current circulation possibly due to earth movements or a shift of the Earth's poles of rotation.

The Ice Age affected erosion and sedimentation all over the planet and had a marked effect on sea-level. When the ice was at its maximum, sea-level fell by at least 100 metres. This caused rivers to cut deeply into their lower courses and across the partially exposed continental shelves. During these times no sea existed between the British Isles and Europe. In the interglacial stages sea-level rose, marine erosion took place at levels above present sea-level, and the lower reaches of rivers became choked with sediment. It is estimated that sea-level would rise about 100 metres if the present day icecaps melted. In addition to sea-level changes, the weight of ice was sufficient to depress the Earth's crust in some places. At the northern end of the Baltic Sea, the crustal surface is rebounding at the rate of 10 millimetres a year following the melting of the Scandinavian icecap.

Another remnant of the Ice Age occurs today in areas bordering the Arctic. For some distance below a glacier, the subsoil and rock is at sub-zero temperatures, and these temperatures persist after the retreat of the ice. Only the top few centimetres of soil thaw each summer. Water cannot drain away, so that solifluction and water-logged ground are widespread. The phenomenon is called *permafrost*. During and soon after the Ice Age it was widespread, but it is now limited to northern Alaska, Canada, and Siberia, where it seriously hampers the development of these regions.

Extent of glaciation in the British Isles in the Quaternary Ice Age.

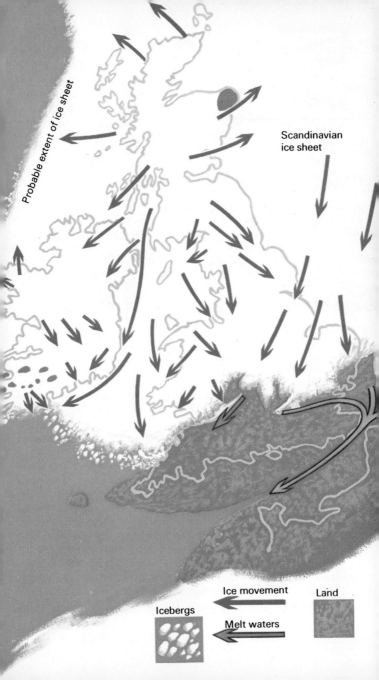

Probable extent of ice sheet

Scandinavian ice sheet

Ice movement

Land

Icebergs

Melt waters

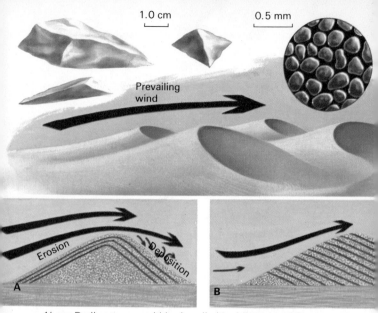

Above Dreikanter are pebbles bevelled by blown sand *Centre* Barchans, which occur in sporadic units or chains, or in colonies *Below* Internal structure of *A* stationary dune *B* migrating dune showing the formation of cross-bedding *Inset* Millet seed sand grains.

Wind

Regions receiving less than 250 mm of precipitation each year are deserts. Deserts may be hot or cold. Weathering is principally by disintegration, and wind is the main erosive force. It removes weathering products, and may leave bare rock or expanses of pebbles.

Wind can erode great depressions such as in western Egypt where erosion has reached below sea-level, only ceasing to erode more deeply because the water-table has been reached. Such features are called deflation hollows. Armed with sand grains, the wind erodes by sand-blasting. Much of the erosion is just above ground level, undercutting and smoothing steep faces and giving a characteristic shape to pebbles which are called *ventifacts* or *dreikanter*. Sand grains blown by the wind are rapidly sorted and rounded; the latter gives the 'millet seed' shape characteristic of many desert sands.

Sands can be blown over hundreds of kilometres, and large parts of deserts are covered by dunes. These move as the wind blows the sand particles. The shapes of dunes are very varied, but a common type is the crescent-shaped *barchan* illustrated here. In the Sahara some dunes of this type are over 300 m high. Dust storms can carry the finest particles out of the desert region to be deposited as *loess*.

Wind is also an important agent in areas bordering glaciated regions where there is little free water. During the Quaternary glaciations, wind-borne sediments were spread over large areas of Eurasia and North America: in Britain these wind-borne sediments are called *brickearths*. Wind erosion and transport can also be important around large beaches where the prevailing winds are on-shore.

Wind erosion is associated with deserts, but water erosion remains important. Rain is rare, but when it falls it does so as intense rainstorms. The lack of vegetation allows vigorous erosion by the short-lived streams. Vast amounts of detritus are carried down *wadis* – steep sided, normally dry valleys – to be deposited as vast alluvial cones in low lying areas. Once the rain stops, waters evaporate to give rise to salt deposits called *salinas*, or they rapidly soak into the sands and underlying rocks allowing the wind to act again.

Desert conditions: *Left* A wadi which fills with flowing water during infrequent rainstorms *Right* Alluvial fan and desert with old channels and salt deposits.

The sea

The sea erodes and transports by means of waves and tides. Waves are created by wind blowing on the sea surface. Their size depends upon wind strength, duration of wind action, and *fetch*, which is the distance over which the wind blew on the waves. In the open ocean only the wave-form progresses, while the water has an oscillatory motion. When the sea shallows to depths less than the distance between wave crests – their *wavelength* – the sea floor drags on the waves, the waves close up like cars slowing on a motorway, and become steeper until the wave breaks with a forward rush on to the shore.

Tides are the twice daily movement of great masses of ocean water due to the gravitational pull of the Moon and, to a lesser extent, the Sun. The tidal range, which is the difference between high and low water, is only about one metre in the open ocean but can reach fifteen metres in restricted channels. Twice a month, tides have a larger than average range; in simple terms, these *spring* tides are caused by the combined in-line pull of Sun and Moon. Half way between spring tides are *neap* tides of less than average range when the Sun and Moon pull at right angles. The

Above Wave motion *Centre* The arrows show how waves curve to attack a headland *Below* The effects of wave action.

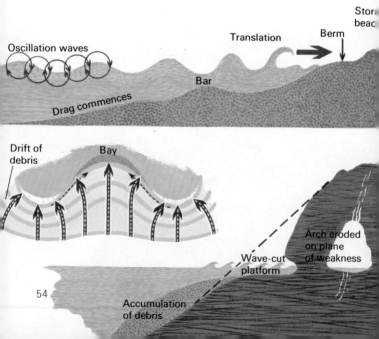

Oscillation waves

Drag commences

Bar

Translation

Berm

Stor beac

Drift of debris

Bay

Wave-cut platform

Arch eroded on plane of weakness

Accumulation of debris

Shoreline scenery.

ebb and flood of tidal waters make strong currents capable of erosion and transport.

Erosion of a cliff by the sea is by *hydraulic* action – waves compressing water and air in cracks; by *corrasion* – the result of pebbles being hurled at the cliff; and by *corrosion* – the solution of rocks by sea water. In addition, rock-boring marine creatures weaken the rock. Wave action breaks down, sorts, and rounds the rock fragments produced by erosion.

The erosion of cliffs creates a *wave-cut platform* which can be seen at low tide. Planes of weakness are attacked to form caves, arches, and finally stacks, which are pillars of rock left on the wave-cut platform. The configuration of a coastline reflects, to a great extent, the hardness of the rocks which comprise it.

Given time, the sea would eventually cease to erode a cliff, because the waves would lose all their energy as they crossed the wave-cut platform. Thus, the shoreline would be graded or in equilibrium. Such shorelines are rare because a number of processes constantly change sea-level.

Erosion products are either accumulated in deeper water or deposited as beaches which, on irregular coastlines, are confined to bays. Beaches are composed of sand, gravel, and cobbles. The particles are constantly moved back and forth by the swash and backwash of breaking waves, and by this action are rounded and sorted. Beaches exist in a state of equilibrium, a balance between wave force, particle size, and beach steepness. Thus, a storm beach can form above high tide level composed of the coarser particles of the beach. Most beaches exhibit nearly flat steps or *berms* which represent high water marks, while just below low water mark, at the limit of backwash, a submerged bank or bar develops.

Most waves break obliquely to a beach causing longshore currents parallel to the shore which are capable of moving sediments. These currents are interrupted by strong rip currents moving from shallow to deeper water. The oblique break of waves on the beach causes *beach drift* by moving sediment laterally. The total effect of the oblique impact of waves is *longshore drift*. When large quantities of sediment are so moved, a bar or spit may develop

Above Longshore drift and the effect of groynes.
Below Estuary with spit and marshes.

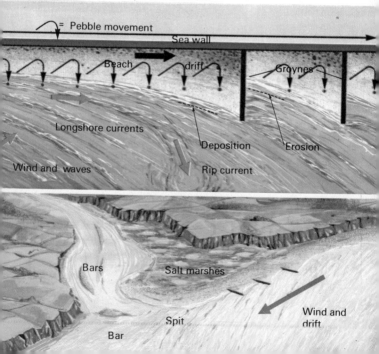

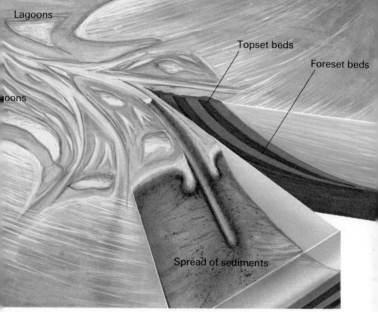

Delta showing distributaries and lagoons.
Inset Section showing simplified structure of delta sediments.

across the mouth of a bay creating a lagoon. In its quiet waters fine sediments including organic matter accumulate to low water mark making tidal flats and salt marshes. These can also border estuaries.

An estuary is the partially submerged mouth of a river. Estuaries may become choked with sediments from river and sea. This may necessitate mechanical dredging if the estuary is to remain navigable. Spits may also block off an estuary.

When the amount of sediment brought downstream exceeds that which can be moved by erosion and longshore drift, a river builds a delta. On entering the sea or lake, a river slows and deposits its load. The sediments and structure of most deltas are complicated. Distributary channels bounded by levees develop, and lagoons and marshes are created. A section of delta shows inclined, coarse *foreset beds* which grade out to fine *bottomset* beds. As the delta grows, the foreset beds are deposited on the bottomset, and some slump under their own weight. Organic-rich sediments are deposited in the lagoons making *topset* beds above the foreset beds.

All over the world, man expresses elevation in terms of height above sea-level. Sea-level and variations in it have been recorded for many years by tide gauges. A study of tide records shows that sea-level varies in response to more factors than simply waves and tides. Low atmospheric pressure causes sea-level to rise either locally in a storm or regionally in relation to the Earth's climatic belts. Higher temperature also has the same effect by expanding sea water. Prevailing winds can pile water against a shore, and even variations in the gravitational field affect sea-level.

While any movements which cause a change in the configuration of the ocean basins can alter sea-level, the most outstanding world wide (*eustatic*) changes resulted from the advance and retreat of the ice sheets in the recent Ice Age. Locally, sea-level is affected by vertical movements of land areas. In addition to movements resulting from glacial rebound (see page 50), there are slow regional, vertical movements of the continental crust. Such *epeirogenic* movements cause the emergence or submergence of continental areas: a careful study of tidal records reveals that the eastern part of the British Isles is sinking while the western part is rising at the rate of about a millimetre a year. Spectacular sea-level changes occur in regions of crustal instability, such as movements on planes of weakness (faults), or associated with mountain building

The Temple of Serapis, originally built on dry land, was then almost completely submerged as is indicated by the borings of marine molluscs, but it is now above water again as the sea-level has dramatically changed.

Above A typical raised beach. *Below left* A submergent coastline. *Below right* An emergent coastline.

(*orogenic*) movements. Most mountains, even Mount Everest, are composed of sedimentary rocks deposited below sea-level. Many examples of sea-level changes occur in volcanic areas: a famous instance is the Temple of Serapis, near Naples; built in the 200s A.D., totally submerged by the 1400s, it is now nearly fully emerged again.

Clearly stable, mature coastlines are rare. The development of any coastline during sea-level changes will depend upon rock hardness and the topography of the land. Coastlines of emergence may show raised beaches, rejuvenated valleys and areas of exposed sea floor. Hilly coastlines subjected to submergence will have drowned valleys, called *rias*, and many headlands. If the grain of the country runs like fingers into the sea, it is called an Atlantic coast; the grain of a Pacific coast is parallel to the coastline.

Given time, a straight coastline will be produced, partly by advance due to deposition, but mainly by erosional retreat. Most coastlines are complex, however, because before maturity is attained, sea-level changes.

Left Lake in a temperate region. *Right* Lake in a desert region.

Lakes

Lakes are landlocked bodies of water which range in size from small pools to large inland seas like the Caspian Sea. Lakes drained by a river are usually fresh; those with no outlet are often saline. Geologically speaking they are temporary features because lacustrine (lake) sedimentation is often rapid and lake barriers are usually quickly eroded: all that remains is a flat, often marshy area. In humid regions lake sediments are usually highly organic, although varved sediments may exist in glaciated areas. In arid regions, salts are precipitated and these are mixed with river and wind-borne sediments. At times such lakes dry out completely to become *playas.*

There are many possible origins for lakes; the most common is the blocking of a valley by a barrier. Additionally, lakes are caused by earth movements, by erosional overdeepening by solution making hollows in soluble rock, and by volcanic action.

Morainic barriers block drainage in many areas recently covered by ice, and glacier ice still blocks valleys in Greenland and the Alps. Alluvial fans and landslides can also block valleys; in the latter case the rapid erosion of debris can cause disastrous flooding downstream. The raising of a river above its flood plain by natural levees can cause lakes by preventing tributary streams joining the main river, while oxbow lakes are formed by the abandonment of a meander loop. Near the sea, lakes can form between elevated distributary channels of a delta or behind spits and beaches.

Earth movements can cause lakes: the Caspian Sea was created by epeirogenic movements, and Lake Victoria and Lake Eyre in Australia by the marginal uplift of basins. The tilting of a large area can reverse river flow to make a lake, such as Lake Kioga in East Africa, while mountain building can cause lakes in intermontane depressions. Faulting, the movement of large blocks of the Earth's crust in relation to each other, has caused many lakes, including many in the East African Rift Valley.

Overdeepening by glacial action creates natural lake basins, such as the corrie and valley lakes of the English Lake District, and the Great Lakes of North America, produced by continental icesheet scour. Wind erosion also causes depressions which may become water-filled. In volcanic regions lava-flows can form lake barriers, while in quiet times lakes often develop in craters and calderas.

East African Rift System showing lakes, rivers and volcanoes.

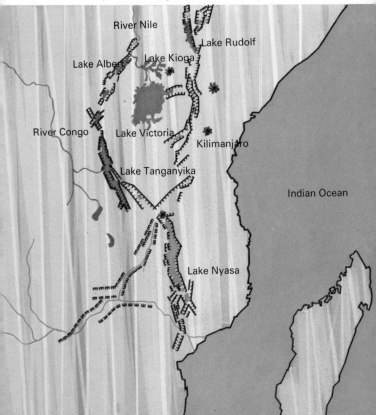

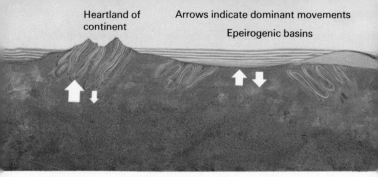

Heartland of continent

Arrows indicate dominant movements

Epeirogenic basins

Very simplified pattern of provinces of sedimentation with arrows showing dominant directions of crustal movements.

SEDIMENTARY GEOLOGY

Provinces of sedimentation

The products of weathering are transported as clastic (broken) particles or in solution. Sooner or later sedimentation by deposition and chemical or biological precipitation occurs. Sedimentation may be either short-lived, being rapidly succeeded by further erosion and solution, or continuous over a wide area for prolonged

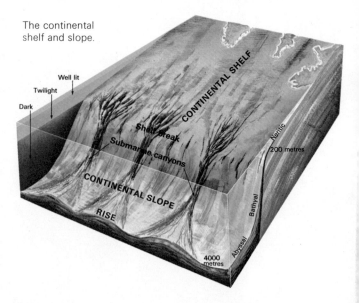

The continental shelf and slope.

Well lit

Twilight

Dark

CONTINENTAL SHELF

Shelf Break

Submarine canyons

Neritic

200 metres

CONTINENTAL SLOPE

RISE

Bathyal

Abyssal

4000 metres

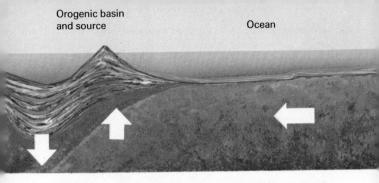

Orogenic basin and source

Ocean

periods leading to thick accumulations of sedimentary rocks. There are three broad sedimentary environments: continental, mixed (coastal), and marine. In the continental environment, sediments may be deposited subaerially (on land) or subaqueously (under water). These sediments are usually clastic and are often strongly oxidized, exhibiting a red coloration. Sediments of the coastal environment are often deposited under brackish conditions. They include the varied sediments of deltas, estuaries, and coastal lagoons, together with those beach sediments deposited above tide level. Sediments of the marine environment are grouped in three broad categories: *littoral* (between tides), *neritic* (in depths up to 200 metres), and *deep water*. Often coarsely clastic near land, marine sediments are finer in deeper water. Organically formed sediments, frequently high in calcium carbonate content, may be abundant, particularly when land derived (terrigenous) sediments are sparse. Sediments are transported by wave action, tides, ocean currents, and turbidity currents.

The marine environment

1 *The continental shelves* These are the submerged edges of the continents and range in width from several to hundreds of kilometres. Light can reach the sea floor which rarely exceeds 200 metres in depth. Marine organisms, particularly those creatures which live on the sea floor – the *benthos* – abound, contributing to the wide variety of sediments. In the clear waters of warm tropic seas corals form thriving reefs. Much of the sediment is constantly moved by waves and tides and there are many patches of bare rock. Sediment accumulations generally occur where the shelf surface is sinking. The surface, inclined about 1:1000 oceanwards, is nearly flat, possibly due to erosion at times of low Ice Age sea-levels. Submerged valleys, rocky shoals, and moving sand-waves which resemble

63

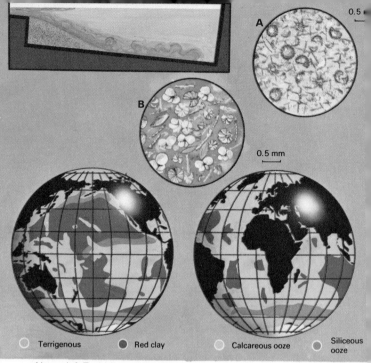

Above left Turbidity current in an experimental tank.
Above right and centre Deep sea oozes: *A* radiolarian (siliceous)
B Globigerina (calcareous).
Below Distribution of ocean sediments.

Terrigenous   Red clay   Calcareous ooze   Siliceous ooze

dunes, are the main features. Continental shelves comprise about 8 per cent of the Earth's surface area.

2 *The continental slopes* At the *shelf break*, the outer edge of the continental shelf, the continental slope falls away steeply from about 200 metres to about 4000 metres with an average slope of 1 in 15. The slope is rarely smooth and is often cut by valley-like *submarine canyons*. Some canyons cross the shelf from the mouth of a large river, but most start at the shelf break. They are thought to have been eroded at times of lower sea-level by turbidity currents, subaqueous flows of sediment-laden water which move downslope because they are denser than the surrounding waters.

3 *The deep ocean* At the foot of the canyons fans of sediment spread out, and the slope gives way to the *continental rise* which is composed of a thick pile of sediments. Beyond, at nearly 5000

metres depth, are the *abyssal plains* with scattered abyssal hills; sedimentation there is slower than on the rise. A prominent feature on all ocean floors is the world-encircling network of *oceanic ridges* which are often 2000 kilometres across. Much shallower than the abyssal plains, they rise to a central ridge which is often cut by a median rift valley. Fault scarps (see later) also cut the ocean floor. In the western Pacific and parts of other oceans, volcanic *island arcs* abound; these are often paralleled by deep *trenches* as much as 11 000 metres deep – *hadal depths*. Other ocean features include scattered volcanic islands, some associated with the ridges, *coral atolls* (see illustration), and submerged flat-topped islands, known as *guyots*.

The sediments of the slope and ocean are varied. The amount of terrigenous material decreases away from land, although there are some local sources of volcanic detritus. Calcareous ooze, consisting of the microscopic shells of the planktonic (surface floating), single-celled creature, *Globigerina*, covers large areas of ocean floor. At depths greater than 4000 metres, these shells are dissolved. In other areas there are siliceous oozes made from siliceous plank-tonic radiolaria and diatoms. Silica is more resistant to solution and can accumulate at greater depths. Elsewhere, fine insoluble red clay abounds, often containing manganese nodules which have a potential economic value for copper, nickel, cobalt, and manganese. In more polar regions, sediments dropped from melting icebergs add to the variety of sea floor sediments.

A possible evolution of atolls caused by the progressive sinking of a volcanic island.

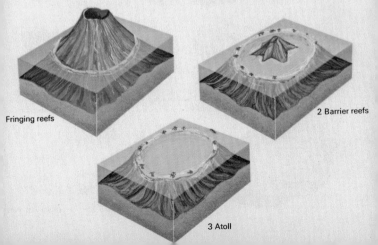

Fringing reefs

2 Barrier reefs

3 Atoll

Sediments and sedimentary rocks

The distinction between sediment and sedimentary rock is arbitrary; the change is brought about by cementation which is the precipitation of minerals from water trapped at the time of sedimentation (connate water) and ground water, and compaction. The process is called lithifaction. Accumulations vary in thickness from a few metres to nearly twenty kilometres, and although they comprise only 5 per cent of the crustal volume, sedimentary rocks cover about 75 per cent of the Earth's land surfaces.

Most sequences of sedimentary rock are characterized by distinct layering, called bedding or stratification, caused by pauses in sedimentation or compositional variations. Sedimentary rocks often contain evidence of their source rock, of the method of transport, and of depositional conditions; in addition, fossils which can be used for dating are restricted to them.

Sediments are grouped into three categories: *clastic* or fragmental; *organic*, composed of plant and animal remains; and *chemical*, produced by direct precipitation. Clastic rocks are subdivided by particle size into *rudaceous* (pebbly), *arenaceous* (sandy), and *argillaceous* (clayey), while further divisions depend on particle

Common sedimentary rocks.

Calcareous shale

Lamin shale

Fine sandstone

Iron stained sandstone

Arkose

Chalk with flint

shape and composition. Few sedimentary rocks are well sorted, and most are mixtures requiring names like 'silty sandstone'.

The rudaceous rocks include conglomerates and breccias, with rounded and angular fragments respectively. Arenaceous rocks are composed mainly of particles between two and one-sixteenth millimetres in diameter. The many types reflect variations in mineral composition and cement: quartzose sandstone, composed mainly of quartz grains; ferruginous sandstone, cemented by iron oxides; arkose, with a granite-like composition; greywacke, a mixture of rock fragments, feldspar, and clay. Argillaceous rocks include clays, mudstones, and shales. Organic rocks include shelly, coral, and algal limestones which are predominantly composed of calcium carbonate, and each is named after its principal component. Peat, lignite, and coal are all composed of vegetable matter. Chemical rocks include ironstones, dolomite which contains magnesium as well as calcium carbonate, and a range of carbonates, sulphates (gypsum and anhydrite), and chlorides (including halite – common salt) produced by evaporation and collectively called evaporites. Limestone can occur in each category, produced organically or chemically, while subsequent erosion can lead to clastic limestones.

Oolitic limestone

Breccia

Conglomerate

Shelly limestone

Calcareous tufa

Crystalline limestone

The basic sedimentological unit, readily observed in most exposures of sedimentary rocks, is the bed. Its roughly parallel upper and lower boundaries are called bedding planes. Few beds are uniform throughout, and many exhibit internal divisions called laminations which are created by slight variations in grain size. Deposition of a bed from a fast-flowing current can cause current bedding. In other beds, there is an upward decrease in particle size giving rise to graded bedding.

A wide variety of structures can occur on the underside of beds including the casts of organic tracks and trails, current-eroded flute marks, and tool marks caused by the impact of current-transported particles on the depositional surface. The upper bedding planes of beds may show mud cracks, caused by drying and contraction, salt crystal shapes and raindrop marks, all indicating subaerial rather than subaqueous conditions. Fossil footprints are indicative of very shallow water conditions. Ripple marks can be asymmetrical or symmetrical, formed by wind or water currents, or by gentle forward and backward water motion respectively. An example of a larger scale sedimentary structure is the washout, formed by channel-

Some common sedimentary structures.
Inset Two interpretations of a structure. The correct interpretation could be derived from study of sedimentary structures.

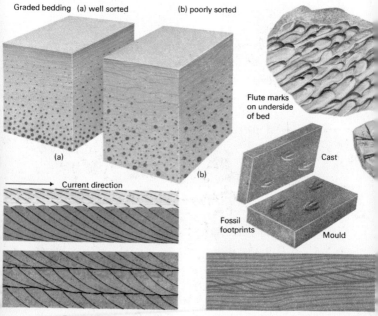

Graded bedding (a) well sorted   (b) poorly sorted

(a)

(b)

→ Current direction

Flute marks on underside of bed

Cast

Fossil footprints

Mould

Current bedding

scour and subsequent infill.

Sedimentary structures provide three kinds of useful evidence: they inform about depositional conditions; some permit the deduction of ancient patterns of wind and water current directions; and they aid in unravelling the order of succession in areas of complicated rock deformation by indicating the top and bottom of beds.

All the structures described above are *primary* structures, that is, they were formed at the same time as the sediment was deposited. Sedimentary rocks also contain a wide variety of *secondary* structures sometimes formed long after deposition by mineral precipitation from connate waters and circulating groundwater. Concretions, geodes (hollow, crystalline, globular masses) and flint nodules (hard, dark, irregular masses of silica deposited in chalk) are three examples.

Grain size, grain sorting, particle shape, and roundness are all part of the texture. Textures, like structures, provide information about conditions before, during, and after deposition; they are, however, best studied when magnified by lens or microscope. Many limestones have distinctive textures such as the oolitic texture (small spherical particles created by calcium carbonate precipitation in warm shallow seas) which is abundant in limestones of all ages.

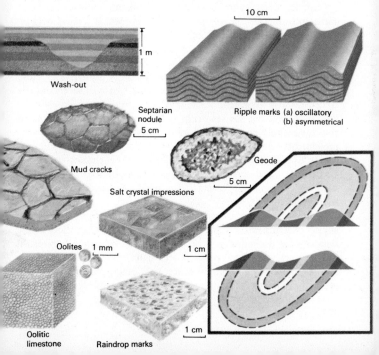

1 m

Wash-out

10 cm

Ripple marks (a) oscillatory
(b) asymmetrical

Septarian nodule
5 cm

Mud cracks

Geode
5 cm

Salt crystal impressions

1 cm

Oolites 1 mm

Oolitic limestone

Raindrop marks
1 cm

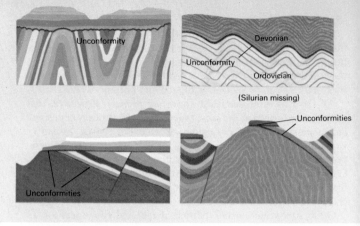

Some types of unconformity.

Earth movements and climatic conditions exert strong controls over the type of sediment deposited. A sedimentary rock is the combined product of source material, weathering, transport, depositional conditions, and post-depositional changes.

Climate exerts controls over weathering and transport, while earth movements control the speed of events. If a land area is sharply elevated and the adjacent sea floor subsides, then particles will move rapidly from source to place of deposition. The resulting sediment will probably be texturally and mineralogically *immature*, that is, the grains will be angular and poorly sorted, and those minerals which would normally be mechanically and chemically destroyed will be present. If the sea floor subsided very slowly there would be time for the reworking of the sediments by currents so that they would be *mature* before burial. Clearly the types of earth movements which occur are important. Gentle, epeirogenic movements usually give rise to mature sediments including quartzose sandstones, and organic limestones. Pronounced movements, such as orogenic movements, can produce immature sediments such as arkoses and greywackes.

Sedimentary successions can be thick or thin, but an abundant supply of sediments must be matched by a sinking of the floor of the depositional basin if a thick accumulation of sediments is to result. Sea-level changes are common and cause marine *transgressions* and *regressions*, the submergence and emergence of land respectively. During a transgression shoreline sediments and progressively deeper water sediments successively *overlap*, while during a regression the converse occurs. A succession of transgressions and regressions can lead to cycles of sedimentation.

The shoreline sediments of a transgression make a basal conglomerate for the new succession. This conglomerate is usually associated with an *unconformity* which represents a break in time or in the succession of rocks. If pronounced earth movements occurred before the transgression there will be an angular unconformity between the successions.

Sedimentary processes can lead to the concentration into economically valuable deposits of otherwise scattered particles. Tin, gold, diamonds, and titanium minerals are concentrated by rivers and/or wave action into *placer* deposits. Coal is a sedimentary rock deposited under special cyclical conditions, while oil accumulates in certain types of sedimentary rocks. Common salt, potash, and similar minerals are sedimentary rocks precipitated due to evaporation under special climatic and tectonic conditions.

As the land is submerged, deeper water sediments overlap. When the land emerges the sediments become shallow water in type, and cycles of sedimentation may occur.

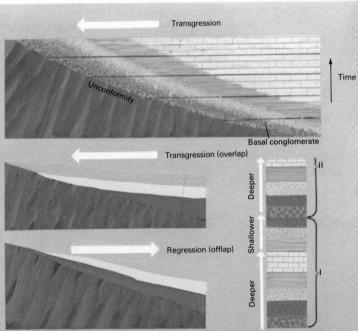

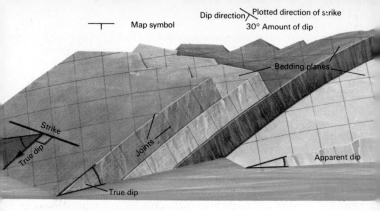

Dip direction — Plotted direction of strike
30° Amount of dip
Map symbol
Bedding planes
Strike
True dip
Joints
True dip
Apparent dip

STRUCTURAL GEOLOGY

Sedimentary rocks are usually deposited in almost horizontal beds, yet when seen in outcrop they are often inclined, folded, or offset. While some distortions are caused by surface (*non-diastrophic*) movements such as slumping or ice pressure, most structures are brought about by crustal (*diastrophic* or *tectonic*) movements. Beneath a superincumbent (overlying) pile of sediments beds may deform without fracturing; in other circumstances fracturing may accompany deformation.

Rock deformation can be studied by measuring the angle of *dip* which is the maximum inclination from the horizontal, and the dip-direction of a bedding plane. These are determined using a clinometer and compass, and the information is usually plotted on a map using appropriate symbols. The *strike* of a bedding plane is horizontal so that it is at right angles to the dip direction. *Strike-lines* which are lines drawn parallel to the strike, can be used to contour a bedding plane. Straight, parallel, equidistant strike-lines at equal vertical intervals indicate an evenly inclined plane.

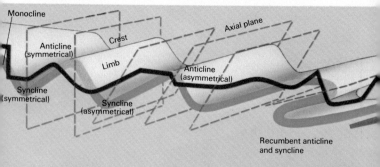

Monocline

Anticline
(symmetrical)

Crest

Axial plane

Limb

Anticline
(asymmetrical)

Syncline
(symmetrical)

Syncline
(asymmetrical)

Recumbent anticline
and syncline

Fold structures.

Dip, strike, and cleavage.

Angle of dip

Simple clinometer

Fracture cleavage in fold

Folded beds are the result of lateral compression. The simplest folds are symmetrical anticlines and synclines. *Anticlines* are arch-like, the beds comprising the limbs dip away from the crest: when eroded the oldest rocks are seen in the core. *Synclines* are trough-like, the limbs dipping towards each other. The plane which bisects the angle between the two limbs is called the axial plane. The fold axis is the line of intersection of the axial plane and any bed of the fold. The axis may be horizontal or inclined. In the former the strike-lines of any bed are parallel to the axis, while in the latter this is no longer the case and the fold plunges.

A *dome* is an anticlinal structure without an axis because the beds dip away from a centre and the strike-lines are roughly circular. The opposite of a dome is a *basin*.

A variety of smaller structures may be seen in any outcrop. *Joints* are fractures normal to the bedding showing no relative movement, and they seem to develop with the release of burial pressure although they are caused by lateral compression. *Fracture*

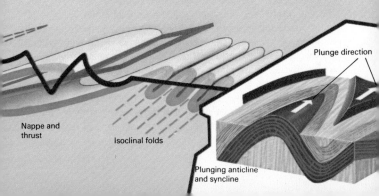

Nappe and thrust

Isoclinal folds

Plunge direction

Plunging anticline and syncline

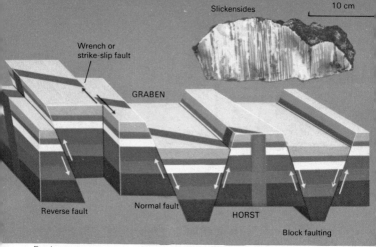

Slickensides

10 cm

Wrench or
strike-slip fault

GRABEN

Reverse fault

Normal fault

HORST

Block faulting

Fault structures.

cleavage results from the breaking of weaker, incompetent beds during folding. The cleavage planes are often roughly parallel to the axial plane. Beds which do not break during folding are said to be competent.

Faulting occurs when rocks fracture and suffer displacement; such movements are the cause of *earthquakes* (see page 84). The displacement may be vertical and/or horizontal. The amount of displacement in any single event is rarely more than a few metres, often much less. Faults can remain active over millions of years, however, and the cumulative displacement in some types of faults may be measured in hundreds of kilometres. Faults are rarely single plane movements; normally they occur as movements on parallel or subparallel sets of planes. Rocks in the fault zone are broken to form a *fault breccia*, and they may even be crushed to a gritty clay called a *mylonite*. Some rock surfaces are striated by the movement; such striations are called *slickensides*.

Normal faults are tensional, the name originated from coal-mining experience in northern England where most faults in mines had this sense of movement. *Reversed* faults are compressional and have the opposite sense movement. Normal faults usually have a steeply dipping fault plane, reversed faults are less steep. The vertical movement is called the *throw*, the horizontal movement is the *heave*. *Thrusts* have the same sense of movement as reversed faults, although the thrust plane may be nearly flat-lying. Faults with marked horizontal displacements and near-vertical fault planes are called *tear*, *transcurrent*, *wrench*, or *strike-slip* faults. They may be called sinistral or dextral depending whether the movement is to the left or right when viewed from across the fault. The San Andreas fault of California, North America is a dextral tear or, more correctly, *transform* fault (see page 81) with a displacement over 600 kilometres. The surface expression of a fault can be pronounced (see page 85), but erosion soon wears the features away. Geologists can often infer the presence of a fault when mapping by the occurrence of 'non-matching' sequences. Faults often aid the flow of groundwater to the surface, and springs can help to locate faults when making a geological map. Faults, too, may be subsequently mineralized because they offer relatively easy passage for ore-bearing fluids.

Combinations of normal and reversed faults can lead to blocks of rock being moved up or down creating *horsts* and *grabens* respectively. The Rift Valley of East Africa is a graben, and is part of a fault system stretching from South Africa to Syria.

Principal structures in England, Scotland, and Wales.

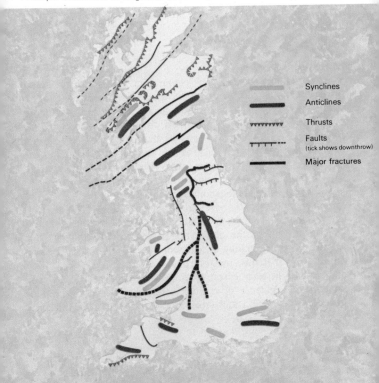

Synclines

Anticlines

Thrusts

Faults
(tick shows downthrow)

Major fractures

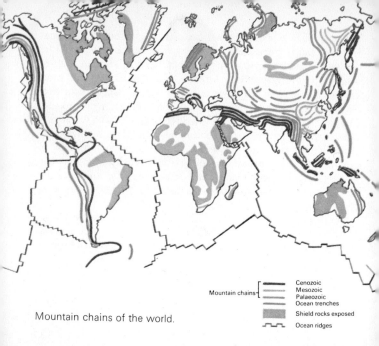

Mountain chains of the world.

Mountains

Most mountains occur in long chains. The longest are the ocean ridges rising over 3000 metres above the abyssal plains.

In some ways the simplest mountain is the *volcano* built up as a pile of ash and/or lava; Mauna Loa, for example, standing 9000 metres above the Pacific floor is of this type and is the largest mountain of all. Others, such as the mountains of Scotland are plateaux uplifted by epeirogenic movements and cut by erosion. Most mountains are the result of tectonic forces, however.

Block faulting made the mountains of the Black Forest, but these cannot compare with the vast fold-mountain chains such as the Alps and Himalayas. All these mountains have the following in common: linear distribution; the chains are narrow compared with their length; composed mainly of sedimentary rocks deposited in structurally formed basins; metamorphic rocks and granite batholiths are exposed in their cores; they are often intensely folded and faulted with evidence of crustal shortening; evidence of a long history of deposition and erosion, and they rest on a thickened prism of *sialic* material.

It is possible to describe a general sequence of the mountain building process, the *orogeny*, which spans tens of millions of years. Sediment as much as twelve kilometres thick gradually accumulates in an elongate, downwarped, *geosynclinal* basin. The waters of this basin need not be deep, but its floor must keep sinking to accommodate the sediments. In the deeper water the sediments, often of greywacke type, are deposited in vast quantities from turbidity currents to give rise to *flysch*. The sediments which fill the geosyncline come from the continents, volcanic islands, and ridges pushed up from the basin floor. Folding starts in the early stages of the geosyncline and is continued, often becoming more intense as deposition and subsidence progress. The lowest sediments are subject to great pressures and heat, and are progressively metamorphosed: from shale to slate, schist, gneiss, and finally to granite. At this time the pressures are intense and uplift begins to be widespread, and associated erosion leads to varied marine and non-marine sediments, called *molasse*, which are deposited around the mountains.

Modern concept of mountain building by plate tectonics *Inset* Traditional concept of mountain building by crustal downwarp and squeezing.

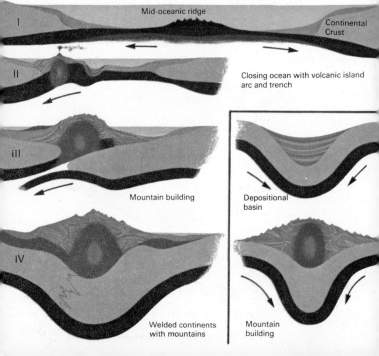

I Mid-oceanic ridge — Continental Crust

II Closing ocean with volcanic island arc and trench

III Mountain building — Depositional basin

IV Welded continents with mountains — Mountain building

The total crustal shortening during an orogeny may be 100 to 150 kilometres, and some rocks are folded and thrust over other rocks for many kilometres in huge *nappes*. The mountains will rise as they are eroded because they are supported by a granitic root, rather like an iceberg, and they must remain in *isostatic* balance, so that as the top is eroded the root pushes the whole mass upwards.

Not all mountains are the same age. The highest mountain chains, the Alps, Himalayas, and Andes are the youngest, because earlier mountain chains have long since been worn away. The mountains of North Wales, Scotland, and Scandinavia are of this type – ancient, once high fold mountains, worn flat and only more recently gently uplifted to be eroded by water and ice.

Oceanic ridges and their associated transverse fault scarps have a different origin. As can be seen in Iceland, where the ridge appears above sea-level, the ridges consist of piled up basaltic lava flows and volcanoes. It is only away from the central ridge with its median rift

Generalized sections across *A* mid-oceanic ridge *B* folded mountain chain (showing effects of increasing metamorphism).

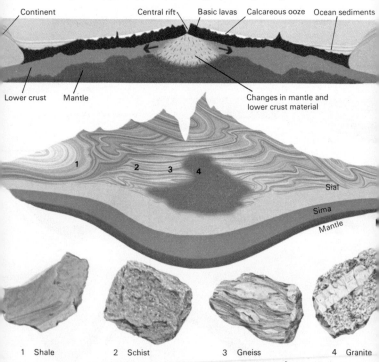

Continent   Central rift   Basic lavas   Calcareous ooze   Ocean sediments

Lower crust   Mantle

Changes in mantle and lower crust material

Sial

Sima

Mantle

1 Shale   2 Schist   3 Gneiss   4 Granite

Increasing grade of metamorphism

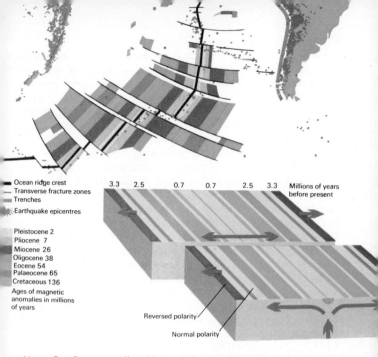

3.3   2.5   0.7   0.7   2.5   3.3   Millions of years
                                            before present

Reversed polarity

Normal polarity

Above Sea floor spreading. Note relationship between earthquake epicentres and major tectonic features. *Below* Symmetrical spreading from ocean ridge crest and polarity reversals in magnetic record.

that ocean sediments are deposited on these igneous rocks. Unlike the fold mountains, which are to a great extent due to compression, ocean ridges form in regions of crustal tension. In searching for an explanation for both of these major mountain systems, a third mountain-like feature must be considered – the strings of islands which comprise the *island arcs* of the Pacific and Caribbean. These islands, often volcanic, are paralleled by elongate ocean deeps called *trenches*, which may be as deep as 11 000 metres. The island volcanoes are invariably andesitic in composition – intermediate between the granite rocks of the mountain chains and the basaltic lavas of the ocean ridges. The connection between mountains, ridges, and island arcs becomes more vivid when these features are drawn on one map, and particularly when this map is compared with the map of earthquake epicentres on page 13. In recent years evidence has been accumulated permitting development of a theory of global tectonics based on the concept of moving crustal plates.

Plate tectonics

Central to the plate tectonic theory is the concept of a number of rigid plates covering the Earth's surface, consisting of crust and upper mantle. These plates are in motion relative to one another, and virtually all earthquake, tectonic, and volcanic activity is limited to the plate margins. The margins are constructive, destructive or, when plates slip past each other, conservative. The ocean ridges are the constructive boundaries where magma wells up to the surface. The significance of these ridges stemmed from the discovery of identical parallel stripes of magnetic polarity on each flank of the central rift. Furthermore, when dated, the rocks of the ridges were found to be progressively older away from the crest. This provided evidence of *sea-floor spreading*, and in places this occurs at speeds of twelve centimetres per year. More evidence came from the study of island and seamount chains. One such chain stretches from Hawaii to beyond Midway. The active volcano, Mauna Loa, fixes the apparently stationary subcrustal magma source. The furthest seamount in the chain, once formed at the same source has been carried 3000 kilometres in 100 million years – three centimetres each year.

Destructive boundaries occur at island arcs; here plates are in collision, one plate passing beneath another. In so doing, the sea

Simplified map of the major plates. The arrows show the direction of plate movements assuming Africa to be stationary.

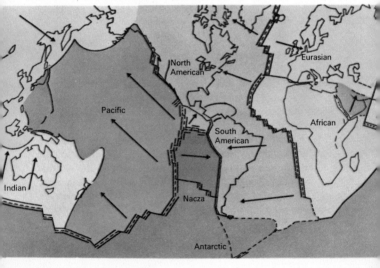

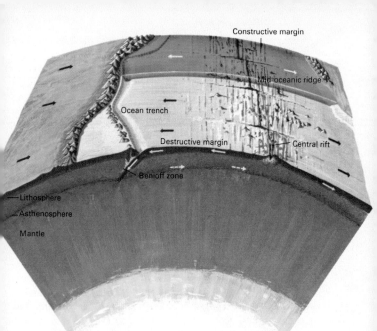

Relationship between plates, plate movements, trenches, ocean ridges, and continents.

floor is pulled down, creating a trench. Earthquake focuses become progressively deeper beneath and beyond the arc along an inclined zone, the Benioff zone; some focuses are over 700 kilometres deep. Frictional and geothermal heat lead to the development of magma sources which erupt to form the islands. A destructive boundary between oceanic and continental crust occurs on the west coast of South America. Again there is a deep trench, progressively deeper earthquake focuses and andesitic volcanoes, and this time there is a mountain range, the Andes. The South American continent is part of a continental and ocean crust of which the eastern limit is the mid-ocean ridge of the South Atlantic. To the east of this ridge Africa is moving away from South America.

Fold mountains like the Alps and Himalayas are also caused by plate collisions at destructive margins. The continents provide sediment, and the mountains are made as one plate is forced beneath the other. Where plates glide past one another at conservative boundaries transform faults are developed, such as the San Andreas fault where the coastal side of the North American continent is

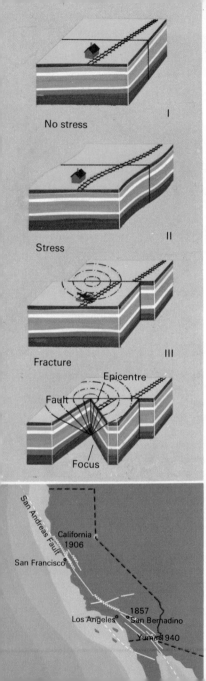

No stress

I

Stress

II

Fracture

III

Epicentre

Fault

Focus

San Andreas Fault

California
1906

San Francisco

1857
Los Angeles   San Bernadino

Yuma 1940

being forced northwards with the Pacific plate.

The theory of plate tectonics offers a scheme which links a great many features present on the Earth's crust. How do the plates move and what drives them? The plates are relatively rigid, and are composed of crust and the top 100 kilometres of mantle. They act together as one unit and are together called the *lithosphere*. Below is the fairly plastic *asthenosphere*. The continents move along with the lithosphere like logs frozen in the ice on a lake. Where the plates move apart magma wells up from the asthenosphere and is extruded as basalt to form new crust, and as it cools it records the direction of the contemporary magnetic field. At collision boundaries, such as island arcs, the lithosphere descends into the asthenosphere.

One likely mechanism for such movements is convection currents in the mantle heated by the core. Some authorities argue in favour of making the lithosphere and top asthenosphere into a closed system, rather like a conveyor belt, with return currents immediately below the moving plates; the whole still

Above Earthquake forces.
Below California fault systems and epicentres of historic earthquakes in America (zigzag lines show breaks in ground during earthquakes).

driven by convection currents. Generally, the deeper system of currents finds wide favour.

When did the movements begin? Presumably they have been in operation since the Earth developed a crust, but the present movements seem to have originated with the breakup of Pangaea, a single supercontinent, about 200 million years ago. By 140 million years ago the Atlantic was beginning to open, while a large ancient ocean, Tethys, between Africa and Europe steadily closed. The southern part of Pangaea, called Gondwanaland, gradually split into South America, Africa, Antarctica, India, and Australia. We can even predict fifty million years into the future: by then Africa will have split along the Rift Valley, the Red Sea will have widened, Australia will be close to Indonesia, the Atlantic will be larger, the Pacific smaller, and North and South America will have drifted away from each other while the State of California will be alongside Alaska!

Pangaea itself was created by earlier movements, and ancient mountain chains which are now worn down, represent earlier collisions of other, older plates. In all these movements of the past and future, continents pass through different climatic belts; indeed, their shapes can affect climatic patterns. This is why coal formed in tropical conditions can be mined in cool areas today and why evidence of ancient glacial episodes can be found in today's tropical regions.

Fault scarp formed during earthquake in Japan.

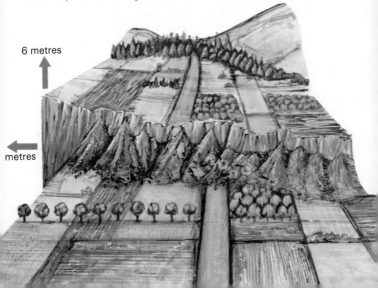

6 metres

metres

Earthquakes

Earthquakes occur when deformed rocks under stress can no longer resist fracturing. The point where the shock starts is called the *focus*, most focuses are within seventy kilometres of the surface but some are nearly 700 kilometres deep. The point at the surface above the focus is called the *epicentre*. The destructive force of earthquakes is well documented, and the table of earthquake intensity illustrates the damage which can be caused. Earthquakes are measured by personal observation of effects and by the magnitude of energy involved, the Mercalli and Richter scales respectively.

There are a number of instances of catastrophic earthquakes such as those that occurred at: San Francisco in 1906 – 8·25; Tokyo, 1923 – 8·2; Chile, 1960 – 8·3 to 8·9; Alaska, 1964 – 8·6.

In addition to the loss of life caused directly by earthquakes, they cause damage in the vicinity of high mountains by triggering rock, mud, and ice avalanches. They can also cause the slumping of submarine sediments, which, if on a large scale, can cause violent water movements creating gigantic waves. These waves can cross the Pacific, the ocean most affected, in twenty-four hours moving at 800 kilometres per hour as well as destroying life on the nearby shore. Not noticeable out at sea because of their great wavelength, they become destructive again on entering shallow water. Such waves, often misnamed 'tidal waves', are called *tsunamis*.

		Mercalli Scale	**Position on Richter Scale**
I	**Instrumental**	– detected only by seismographs	—
II	**Feeble**	– very few people notice	3·5
III	**Slight**	– hardly felt	4·2
IV	**Moderate**	– felt by people walking	4·3
V	**Rather strong**	– sleepers awakened, bells toll	4·8
VI	**Strong**	– trees sway, objects fall	4·9–5·4
VII	**Very strong**	– general alarm, walls crack	5·5–6·1
VIII	**Destructive**	– moving cars affected	6·2
IX	**Ruinous**	– some houses fall, ground cracks	6·9
X	**Disastrous**	– ground opens, landslides	7·0–7·3
XI	**Very disastrous**	– few buildings remain standing	7·4–8·1
XII	**Catastrophic**	– total destruction	8·1 (Max 8·9)

Right Convection currents and moving continents – contrasting theories.
Inset Island arc volcanoes and trench, and their relation with earthquake focuses and the Benioff zone.

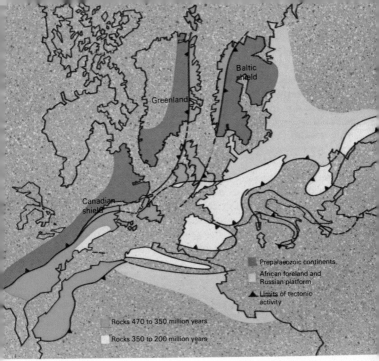

Prepalaeozoic continents
African foreland and
Russian platform
▲ Limits of tectonic
activity

Rocks 470 to 350 million years
Rocks 350 to 200 million years

Above Matching across the Atlantic. Note the size of the ancient Mediterranean.

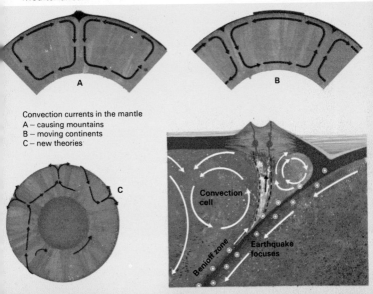

Convection currents in the mantle
A – causing mountains
B – moving continents
C – new theories

Convection cell

Benioff zone

Earthquake focuses

GEOLOGICAL TIME AND STRATIGRAPHY

Contemporary estimates suggest that the solar system is 6000 million years old: Earth's oldest dated rocks are about 3900 million years old. How can such time spans be appreciated and how were they estimated? When compared with man's life span, geological processes are slow. A person living in the Alps, for example, would recognize few changes in his lifetime – some boulders may have rolled down a hillside one winter, a stream may have changed its course, yet the fossils in the Alpine rocks are evidence that the rocks which make the Alps were deposited below sea-level. It was observations like these that led James Hutton, in the late 1700s to write that he could see 'no vestige of a beginning, no prospect of an end', and the English geologist, Sir Charles Lyell, in the early 1800s to develop the *principle of uniformitarianism*: namely that, with some reservations, the present is the key to the past. This contrasts with the concept of *catastrophism*, which explained the record of the rocks in terms of successive world-wide destructive

Estimates of the age of the Earth.

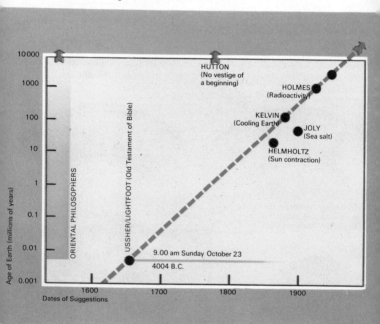

events such as Noah's flood. In the mid-1600s, Archbishop Ussher and a Dr Lightfoot used Biblical evidence to determine that the Creation commenced at 9.00 am on Sunday, October 23 in the year 4004 B.C.; a time span in marked contrast with Hindu estimates of thousands of millions of years! Ussher's dates held sway for over 200 years, but in the late 1800s new scientific estimates of the Earth's age were deduced. Hermann von Helmholtz, Lord Kelvin, and J. Joly, using determinations based on the contraction of the Sun, the heat-loss of the Earth and the salt in the sea respectively, suggested an age of tens of millions of years; Kelvin also made an estimate of 400 million years but withdrew this later. About this time, Samuel Haughton estimated an age of 200 million years, based on sediment accumulation rates. The longer time now accepted is based on the radioactive dating methods.

If we let the 186 metre high Post Office Tower of London represent Earth's age, then truly fossiliferous rocks are represented by the top tenth, and the age of the Alps by the top hundredth of the building, while the time since Christ on this scale is less than the thickness of tissue paper!

Events in the Earth's history (Ages in millions of years).

Phanerozoic (see pages 90/91)

Age of man
Age of mammals
Age of reptiles

First land plants and animals
First animals with backbones
600 my First hard-shelled animals
First soft-bodied animals
First green algae 1000 my

Oldest algal reefs 2000 my

Beginning of build-up of oxygen 2000–2500 my

Oldest fossils 3100 my
Oldest unaltered sedimentary rocks 3300 my

Oldest date of rocks from Earth 3800 my

Greatest age of dated rocks from Moon 4700 my

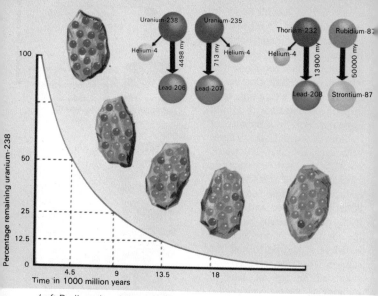

Left Radioactive decay of uranium-238 to lead-206 *Right* Some radioactive elements used in dating rocks giving 'half-life' times.

Radioactive dating

Radioactivity was discovered at about the time scientists were first attempting to estimate the age of Earth by scientific means. By 1913, Arthur Holmes, using dates ascertained from the radioactivity of certain elements, had proposed that the base of the Cambrian system and the beginning of the succession of rocks containing abundant fossils, was 500 million years old. Since that time radioactivity has become the principal method of dating rocks and in 1959, Holmes was able to re-estimate the age of the base of the Cambrian at 600 million years.

This method of dating makes use of the discovery that radioactive elements undergo spontaneous continuous breakdown or decay into stable elements by the loss of particles. By measuring the ratio of original, 'parent', to new, 'daughter', elements, the time which has elapsed since the creation of the parent can be computed. It is usual to refer to the *half-life* period of the decay; that is, the time required for half of a given starting weight of a parent to change into its daughter element. Uranium-238 changes to lead-206 and helium-4 with a half-life of 4498 million years; potassium-40 to argon-40 with a half-life of 11 850 million years; and rubidium-87 to strontium-87 with a half-life as long as 50 000 million years. The carbon-14 dating method using the change of carbon-14 to nitrogen-14 can be used only on organic fragments, and because the half-life is only 5570 years it cannot be reliably used on material older than 50 000 years.

The most reliable dates using radioactive methods are those determined on igneous rocks because their constituent minerals yield a precise starting point for the decay. In metamorphic rocks, the date usually relates to the last episode of metamorphism. Sound geological principles must be applied to relate dated rocks to the rest of the sequence of layered rocks.

The oldest rocks so far dated were collected on the Moon and are 4700 million years old. The oldest dated rocks on the Earth are 3800 million years old; they occur in West Greenland, and are intruded into even older rocks. The earliest fossil bearing rocks are found in South Africa and are 2800 million years old. Many other formations contain early fossils, but the great explosion in life-forms appears to have taken place 600 million years ago at about the beginning of the Cambrian.

Explanation of C14 — N14 cycle.

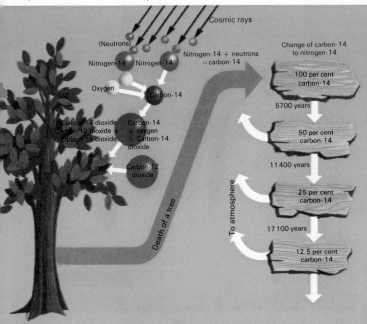

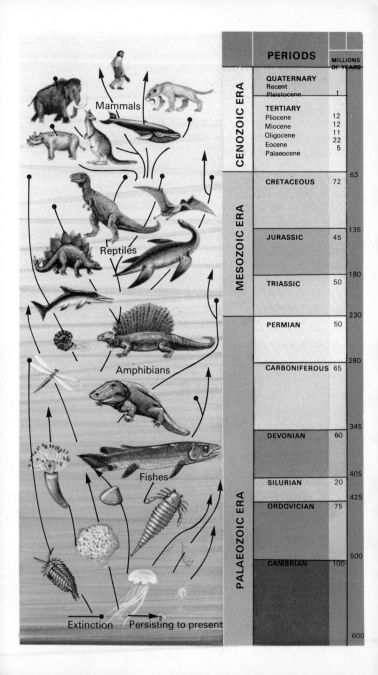

		PERIODS	MILLIONS OF YEARS
CENOZOIC ERA		**QUATERNARY** Recent Pleistocene	1
		TERTIARY Pliocene Miocene Oligocene Eocene Palaeocene	12 12 11 22 5
MESOZOIC ERA		**CRETACEOUS**	72
			63
		JURASSIC	45
			135
		TRIASSIC	50
			180
PALAEOZOIC ERA		**PERMIAN**	50
			230
		CARBONIFEROUS	65
			280
		DEVONIAN	60
			345
		SILURIAN	20
			405
		ORDOVICIAN	75
			425
		CAMBRIAN	100
			500
			600

Mammals

Reptiles

Amphibians

Fishes

Extinction ●   Persisting to present →

Fossils

Fossils are the preserved remains of once living creatures – or the evidence of their existence – found in the succession of rocks. Fossils include completely preserved creatures, carbonized plants, casts of shells, footprints, burrows, and even coprolites which are fossilized excrement.

Completely preserved animals are extremely rare and almost never come from very old strata. Their preservation depends upon special conditions. Woolly mammoths have been preserved in frozen Siberian gravels for 25 000 years, and in several parts of the world animals which fell into tar pits have been fully preserved including the remains of their last meal. Insects trapped in amber (fossil wood-resin) are another form of complete preservation.

Those parts of animals which are very resistant to chemical alteration, such as sharks' teeth, may be preserved unaltered, but in general, fossilization is accompanied by chemical alteration. Circulating waters dissolve chemicals and replace them by others; silica, calcium carbonate, and iron compounds are common replacements. Sometimes the replacement preserves the molecular structure of the original. More often, all traces of the original tissue are lost although the delicate outer form and, in the case of hollow shells, the inner form may be preserved.

Moulds and *casts* of the original creature are another form of fossilization. Here, the original structure has been completely dissolved away leaving a hollow, and because this hollow mould bears the structure of the original it is a fossil. When the mould is filled with fine sediment or minerals carried by circulating waters a cast is created – another form of fossil. Both the inner and outer structure of the original can be preserved in this way.

Very rarely, extremely fine sediments have allowed the preservation of casts of soft tissue as well as the hard parts. Some of the earliest fossils are of primitive algae and bacteria, but fossils of jellyfish and worms have been found in rocks 1000 million years old in Australia.

In addition to the preservation of organisms in one form or another, evidence of organic activity can be fossilized, such as the footprints of dinosaurs or trails of burrowing creatures.

Fossils are used to date and correlate rocks; they provide evidence of conditions at the time when the creatures lived and present a picture of life evolving through time.

The Phanerozoic time scale and some life forms.

Only some of the principal *phyla* (major groupings) of fossils from the Phanerozoic (literally 'obvious life') rocks which are of value to the geologist, can be described in this book. Before the Phanerozoic began, 600 million years ago, life was simple and fossils rare. One explanation for this is that until the Cambrian there was insufficient atmospheric oxygen to permit the varied development of fossilizable animals.

Protozoans are single-celled organisms of which the most important group is the Foraminifera because of its value to petroleum geologists for dating. Much recent ocean sediment is of Foraminifera shells particularly of *Globigerina*. **Coelenterata** include the coral reef-builders, the remains of which comprise significant parts of many limestone formations. **Brachiopoda** are bivalved animals represented today by the lampshell, but were once abundant and rock-forming. **Mollusca** include the Cephalopoda which were once diverse including the extinct ammonites. The ammonites are useful dating fossils. Present day descendants include the nautilus and octopus. Unlike the Cephalopods the Gasteropods have changed little and are represented today by the snail. The Bivalvia are extremely diverse and are represented now by mussels and oysters. **Arthropoda** include the insects,

Some of the fossils which can be found in the geologic time scale.

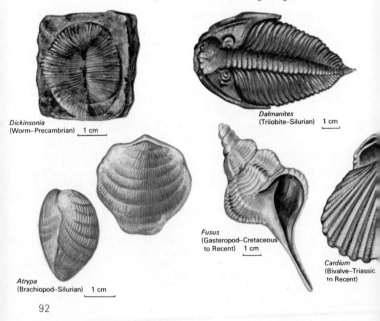

Dickinsonia
(Worm–Precambrian) 1 cm

Dalmanites
(Trilobite–Silurian) 1 cm

Fusus
(Gasteropod–Cretaceous to Recent) 1 cm

Cardium
(Bivalve–Triassic to Recent)

Atrypa
(Brachiopod–Silurian) 1 cm

which though varied and ancient, were not abundant enough to permit rock dating. The extinct trilobites, however, are particularly useful fossils found in Cambrian to Permian rocks. **Echinodermata** are represented today by starfish and echinoids (sea urchins). The echinoids are useful dating fossils for Jurassic and Cretaceous rocks. **Graptolithina** (from Greek meaning 'stone writing') are an extinct group which evolved rapidly in the Ordovician and Silurian, and are useful dating fossils. They are simple in appearance but must have been complex organisms. **Chordata** are the vertebrates – animals with backbones. A diverse assemblage, and while not important for dating rocks, they include fishes, amphibians, and reptiles (including the dinosaurs), birds, mammals, and man. Man is a newcomer, appearing only about two million years ago. **Plants** preceded the animals, are the basis of the food chain, and provide oxygen for the atmosphere. Leaves, seeds, and spores are extremely useful fossils.

In all these groups, there were genera and species which evolved rapidly and took on many forms – these are useful to the geologist for dating rocks. Many groups became extinct, but nearly all types existed for much longer than the time so far occupied by man.

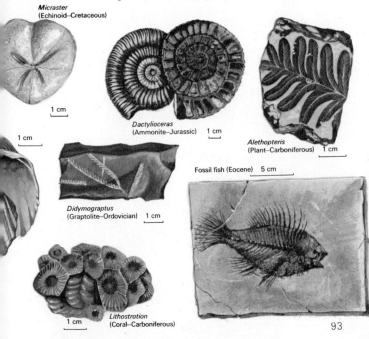

Micraster
(Echinoid–Cretaceous)

1 cm

1 cm

Dactylioceras
(Ammonite–Jurassic) 1 cm

Alethopteris
(Plant–Carboniferous) 1 cm

Didymograptus
(Graptolite–Ordovician) 1 cm

Fossil fish (Eocene) 5 cm

Lithostrotion
(Coral–Carboniferous)

1 cm

93

Stratigraphy

Stratigraphy is the study of layered rocks, their character, correlation, and sequence through time, both locally and on a world-wide basis. Stratigraphy relies heavily on fossils as a method of dating and correlation, so that it is particularly concerned with the Phanerozoic successions. The term 'historical geology', though similar in meaning, has wider implications, and deals with the whole sequence of events in the Earth's history.

The *formation* is a group of beds with common attributes – lithology, structures, and even colour. It can be mapped by 'walking out' the outcrops. It is necessary, however, to relate any formation to other units. The application of the *law of superposition*, propounded by William Smith at the beginning of the 1800s, namely that younger rocks lie on older, is one obvious approach. But Smith's second law, the *law of strata identified by fossils* offers wider possibilities for correlation.

There is no total Phanerozoic succession of strata available in one place for reference; successions are interrupted by breaks and unconformities. Even if such a succession existed it would be of

Time. Time units, rock units, and time-rock units. These figures show how different types and thicknesses of rock can be deposited in equal time intervals.

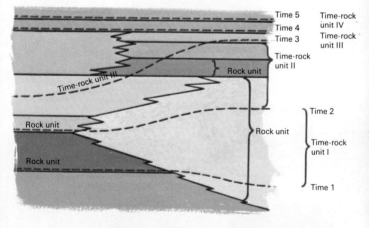

94

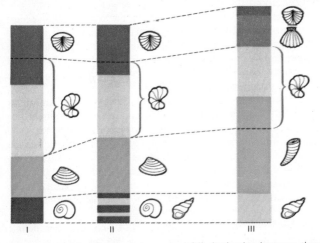

Correlation of fossils. Some types are of limited value because they are restricted by conditions.

limited value because conditions of deposition are never uniform on a regional scale; subaerial, deltaic, clear water organic, and deep water sedimentation all occur simultaneously. Furthermore, these conditions exercise a strong control over the organisms and the fossils which can exist in any place.

The most useful fossils for correlation and dating are of those groups of organisms which, in life, showed wide variations due to rapid evolution, and which were free-floating and independent of water depth and bottom conditions. Ammonites, graptolites, and foraminiferids fulfil these requirements, and consequently are of great value as stratigraphic fossils. Not only single fossil species but associations of different species help the stratigrapher to date and correlate strata.

Once a time scale has been established then the study of variations in *sedimentary facies*, that is the sum of all characters of a sedimentary rock, becomes a meaningful exercise in building up the full picture of past conditions and palaeogeography.

Radioactive dating is now widely used to give absolute dates to the sequence established using fossils. Radioactive dating is the principal dating method when dealing with the Precambrian which, you should remember, represents five-sixths of Earth history since the formation of the earliest dated rocks.

Nearly every rock type and division of the geological column occurs in Britain. Precambrian rocks occur mainly in northern Scotland with some small *inliers* (older rocks surrounded by younger) in England and Wales. Most of these rocks are intensely folded and metamorphosed, and only the Torridonian is wholly sedimentary, composed mainly of sandstones.

The Phanerozoic begins with the Cambrian system (Cambria = Wales) and consists of grits, shales, and slates. The succeeding Ordovician system, consists of greywackes, mudstones, and volcanics. The Silurian system follows, composed of greywackes and mudstones, with limestones in the Welsh borderlands. These rocks were folded during the Caledonian orogeny which ended before the next, Devonian, system. This consists, in Devon, of marine shales and limestones; in Scotland and Wales of sandstones and marls, often called the Old Red Sandstone.

The Carboniferous system follows with limestones, then deltaic grits, and finally the coal measures. The Permian system is represented by limestone and sandstones. All the above systems are grouped into the Palaeozoic (literally 'ancient life') era. The succeeding Triassic system is composed of red marls and sandstones. These deposits and those of the Permian are grouped into the New Red Sandstone although the close of the Variscan orogeny separates these systems.

The Rhaetic is hardly represented, and in Britain it is often grouped with the Jurassic system which includes the marine clays and oolitic limestones of the Cotswolds. These rocks were studied by William Smith, the 'father of stratigraphy' and maker of the first geological maps of Britain. The Cretaceous system starts with deltaic sediments, followed by the Chalk, which gives the system its name. The Trias to Cretaceous rocks are grouped into the Mesozoic (middle life) era.

The succeeding Cenozoic (recent life) era includes the Tertiary systems: the Palaeocene; the Eocene, of which the London Clay is part; the Oligocene, limestones and marls; the Miocene, clays and marls found offshore; and the Pliocene of East Anglia. Their names infer the development of present-day life forms – the *dawn*, *few*, *less*, and *more*, respectively. During the Tertiary extensive volcanic and basalt flow activity affected north-west Britain, and the rocks of southern England were folded. The Quaternary subdivision of the Cenozoic is divided using the same scheme as for the Tertiary, into the Pleistocene which was the time of the Ice Age, and the Holocene which is the short time since the Ice Age.

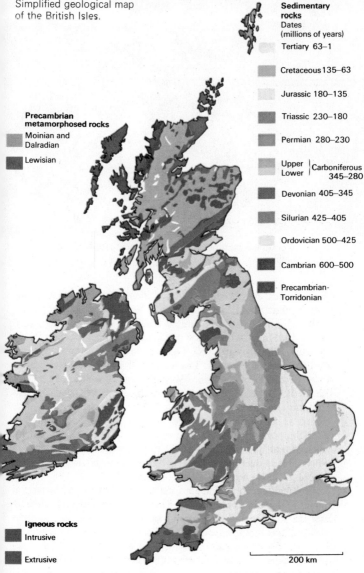

Simplified geological map of the British Isles.

Precambrian metamorphosed rocks
Moinian and Dalradian
Lewisian

Sedimentary rocks
Dates (millions of years)
Tertiary 63–1
Cretaceous 135–63
Jurassic 180–135
Triassic 230–180
Permian 280–230
Upper | Carboniferous
Lower | 345–280
Devonian 405–345
Silurian 425–405
Ordovician 500–425
Cambrian 600–500
Precambrian-Torridonian

Igneous rocks
Intrusive
Extrusive

200 km

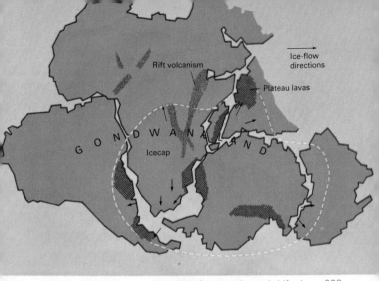

Ice cap of southern continents before continental drift about 280 million years ago.

Episodes of Earth history

The geological evidence for a supercontinent, Pangaea, which started to drift apart about 200 million years ago, leaves little room for doubt. The now widely separated localities in the southern continents and India which show evidence of Upper Carboniferous glaciation can be grouped together into Gondwanaland, the southern part of Pangaea. There they fall within a contemporary Antarctic circle; by the same token, the forests which were to give rise to the Upper Carboniferous coal measures lay near the contemporary equator. The succeeding Permian desert deposits were formed as the continents drifted through the arid zones of that time. Supporting evidence comes from palaeomagnetic studies which reveal ancient poles in different positions from those of today. This is due to the movement of continents after rocks had 'locked in' the contemporary magnetism.

The breaking up of Pangaea caused the division of terrestrial animal and plant communities. Responding to different controls after the split, evolutionary trends led to contrasting communities in spite of a common ancestry. Only free-floating marine organisms, the types that make useful fossils, were unaffected. In each post-split continent a few land creatures existed which showed no evolutionary divergences. This too, is evidence of a common pre-split origin, because it would be unlikely, if not impossible, for exactly similar creatures to have evolved in places separated by thousands of kilometres of ocean.

The Atlantic Ocean opened progressively, and as it opened, Tethys – the ocean between that part of Pangaea called Laurasia and Gondwanaland – seems to have gradually closed causing the Alpine-Himalayan orogeny.

What existed before Pangaea? There were other continental and ocean plates which combined to create Pangaea. The Variscan orogeny which developed between 370 and 260 million years ago affected a belt across Europe and North America, but clearly, there was no Atlantic ocean at that time. Other pieces of the pre-Pangaea jigsaw will be assembled in time. Before that, there were other plates of other configurations. The Caledonian orogeny which started 600 million years ago and ended 370 million years ago was the product of lower Palaeozoic plate movements. Today's ocean floors, which have provided so much evidence about post-Pangaean movements, bear no evidence of earlier events. Only the continents of today contain the records of earlier movements, some far back into the Precambrian.

Climatic belts in Carboniferous times before split of Pangaea.
Inset Glaciated pavement formed in India at this time.

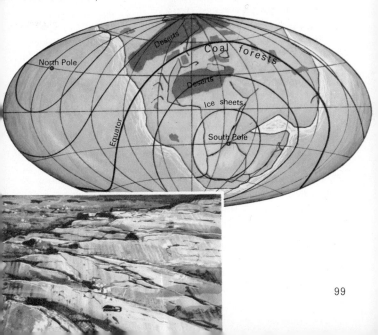

THE ROLE OF THE GEOLOGIST

There are many parts to the integrated whole of geology: mineralogy and petrology – the study of minerals and rocks respectively; palaeontology and sedimentology – the study of fossils and sediments; structural geology interprets the tectonic history of a rock or a mountain chain, and stratigraphy links many aspects together by the correlation and dating of events. In recent times many new fields have developed: geochemistry–earth chemistry; geochronology concerned with radiometric dating; and marine geology. Geophysics is concerned with the interpretation of the Earth's physical condition. Many geologists are investigating the Earth, however, seeking the raw materials necessary for man's well-being, and assisting in engineering enterprises. Much of the remainder of the book is concerned with these applied aspects of geology.

Newspapers often comment upon the 'energy crisis'. Most of the Earth's energy comes from the Sun. If developed, solar energy – the direct use of sunlight – could only provide a small part of our industrial power needs. The best use of sunlight must continue to be the production of food. Wind power, also made by solar energy, can only supply a small fraction of man's energy requirements. Tidal energy is expensive to harness and there are many arguments against it. Hydro-electric power from rivers could provide more energy, but only at the expense of extensive flooding and river diversions. Clearly, the fossil fuels – coal, oil, and natural gas – which are trapped solar energy, will continue to be man's main source of energy, but these fuels are not limitless. Power from fissionable radioactive fuels offers an alternative to past and present solar energy, but uranium and similar fuels are not in limitless supply. Additionally, such fuels present a problem of disposal of their radioactive biproducts. The fusion of the hydrogen atom does not produce harmful wastes, but this form of energy is far from being harnessed. Geo-thermal energy which is heat from the Earth's interior, is another potential source of energy.

Clearly, geologists are going to play an important role in the search to meet the future energy demands. They must also search for the other raw materials which are required by today's industrial societies.

Deep sea drilling ship in operation. Such operations have yielded considerable information about the history of the oceans.

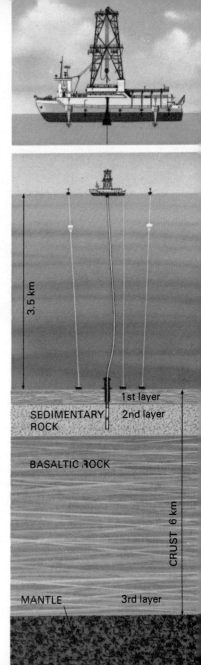

3.5 km

1st layer
SEDIMENTARY ROCK
2nd layer

BASALTIC ROCK

CRUST 6 km

MANTLE
3rd layer

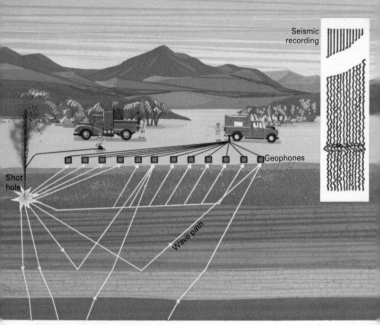

Seismic reflection prospecting.

Exploration methods

Different types of deposits require different methods of exploration, and only the principal methods are described here. The old-time prospector developed an 'eye for country'. He panned streams and followed traces until he found a major placer deposit or the mother lode. The modern geologist also requires an eye for country, and the geological map he prepares is the heart of any investigation. Field work is often supplemented by the study of aerial photographs, and geochemical and geophysical surveys.

Geochemical surveys analyse soils, plants, and stream waters to determine the location of above-average accumulations of minerals. Geophysical prospecting investigates variations in the physical parameters of a region, which may be due to subsurface structures. Seismic surveying makes use of shock waves made by a hammer or explosives which are recorded and analysed. Such a survey may reach only a few metres below the surface to several kilometres depending upon the size of the shock wave produced. Electrical methods measure variations, perhaps caused by an ore body, in naturally occurring electric currents within the Earth, and measure the resistivity to induced electric currents. The latter method can be used to measure the depth to the water-table, or the base of unconsolidated material, and is a useful method for engineering site investigations. Magnetic methods measure distortions in the magnetic field due to the presence of magnetic minerals, and can indicate mineralized fault planes, basic igneous rocks, or iron ore deposits. Gravitational methods use a sensitive spring balance; denser rocks exert a stronger attraction, and this method of surveying shows major regional variations such as the form of a buried basement, folds or salt domes. Radioactive methods use a scintillometer to measure the beta and gamma-ray intensities of rocks.

All these methods are supplementary to the geological map and the study of samples. In marine geology, the preparation of a subsea geological map is not easy, but samples can be recovered by coring while geophysical methods are often more easily carried out at sea.

Test pits and shallow drill holes follow the first investigation, and finally a shaft is sunk or a hole drilled. Should the first 'wildcat' hole be dry, it can still yield further useful geophysical information.

Principle of gravitational surveying: denser rocks cause greater pull on torsion balance than lighter rocks.

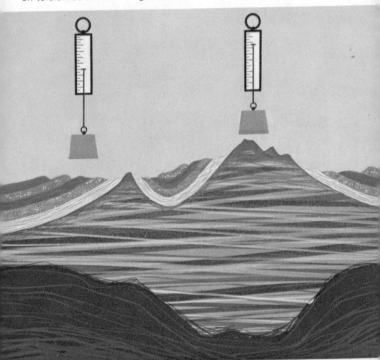

Fossil fuels – coal, oil, and gas.

Life began approximately 3500 million years ago, possibly created from volcanic gases in a lightning charged atmosphere. Living matter is distinguished from mineral matter by a capability of self-reproduction, and by an ability, called *metabolism*, to obtain energy by 'burning' food. Early life was plant-like using solar energy in the form of light to change water and atmospheric carbon dioxide by a process called *photosynthesis* into a range of hydrocarbons such as sugars, cellulose, and starch. Oxygen was liberated in this process. After death, living matter decays by oxidation, reabsorbing the oxygen. If, on death, organic matter is buried in sediment, decay will be halted trapping the hydrocarbons and permitting oxygen to remain in the atmosphere. It was in this way that oxygen became a significant component of Earth's atmosphere. Normally, organic remains in sediment are so scattered that they are hardly discernible, but sometimes circumstances permit the accumulation

Extraction of coal by open cast and underground methods.

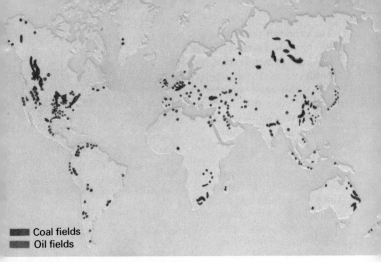

■■■ Coal fields
■■■ Oil fields

Distribution map of coal and oil fields.

of considerable amounts of organic matter. Peat found in boggy land is an example of such an accumulation; in addition, the acid waters of a peat bog inhibit the normal processes of decay.

Ancient peat deposits were buried during the geological past. As sediment accumulated on top of the peat, it lost water and the proportion of carbon in the deposit rose. With sufficient time and pressure brown coal or lignite was formed. The process, carried further, led to the bituminous coals, and these, when affected by intense pressure, were changed into anthracite, a coal consisting of nearly pure carbon. Peat, lignite, bituminous coal, and anthracite form a series of increasing rank: with each successive stage, the gas content falls and the carbon content increases. Conditions particularly suitable for the accumulation of peat-like deposits existed in Upper Carboniferous times when great forests grew just above sea-level, and layers of peat were trapped in cycles of marine and non-marine sediments deposited as the sea transgressed and regressed. Changes in sea-level led to many hundreds of peat layers being trapped, forming the coal seams of the Carboniferous coal measures.

Generally speaking, coal must be mined, and this is a hard and dangerous undertaking, but where seams are flat-lying and near the surface, coal and lignite – which occurs in younger strata than coal – can be recovered by open-cast methods.

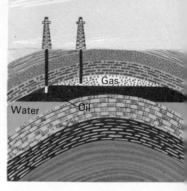
Anticlinal trap

The origin of coal is reasonably obvious, because plant fossils are common in coals, and beneath seams there are *seat earths* which are fossil soils in which there are petrified roots. Coal seams accumulated where the plants grew. In contrast, the origin of oil and natural gas is obscure. It seems that animal and plant remains were trapped in sediments on the sea floor. These remains became deeply buried, and by a combination of pressure and heat, were converted into complex hydrocarbons in solid, liquid, and gaseous states. Oil and gas are not found in the rocks in which they were deposited, but instead they have migrated under pressure into *reservoir* rocks. These are porous rocks which are sealed or capped by impermeable layers. Some of the principal geological traps which permit the accumulation of large quantities of oil and gas to give *oil fields* are illustrated here. The *salt dome* is one of several structures often asssociated with oil fields. The salt, being less dense than surrounding rocks, rises from deeper layers of evaporite sediments causing the strata to dome upwards. Salt is impermeable so that it effectively traps hydrocarbons.

Generally, oil and gas occur

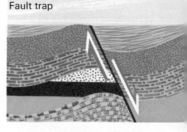
Fault trap

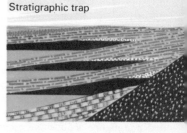
Stratigraphic trap

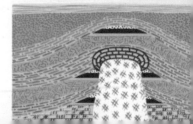
Salt dome

Traps for oil and gas. Note that the gas, oil, and water only fill spaces within rock formations.

in younger strata than those which contain coal seams; this is one reason why the rocks of the continental shelves are proving to be significantly oil-bearing.

Geophysical methods are important in the search for oil-bearing structures, but the only way to prove the existence of oil in economic quantities is to drill. Even with modern exploration methods many holes are dry, but dry holes can provide further geophysical data which can be used for the continued search for oil and gas elsewhere.

Electrical and radioactive logging methods have been generally applied in recent years to aid in the subsurface correlation of oil wells and to a better understanding of the geology of an area where the rocks cannot be examined at first hand.

The first of these makes use of a loop of wire which will induce electric currents in any conductive substances in the rocks as it is lowered down a borehole. The traces of the electromagnetic fields induced are then examined carefully. Plots of resistivity might show a marked increase in resistance for an anhydrite band, for example. Similarly, many radioactivity methods which originally arose from the ever-increasing need for uranium, have been adapted for use in the petroleum industry. As mentioned before, the scintillometer measures both beta and gamma-ray intensities, and the Geiger counter measures gamma-rays only. Even tiny variations in radio-activity can be detected and values may be plotted down a borehole to show how the radioactivity can change with changing lithology. For example, a shale band might give an increase in the gamma-ray intensity, whereas anhydrite would correspond to a distinct trough. A combination of the techniques can produce a 'fingerprint' to positively identify a horizon without ever actually seeing it.

The largest oil fields are in the Middle East, but important fields exist in Venezuela, Nigeria, North America, and Russia. Significant discoveries have been made in the North Sea and Alaska, but in both these areas exploration and extraction are difficult and expensive. More oil fields will be found, and some oil can be extracted from *oil shales* and *tar sands*. Nevertheless, the demand for oil and gas is now so great that serious shortages will probably occur within the next two decades. Alternative sources of energy are coal, the considerable reserves of which can last much longer than those of oil, and nuclear fuels.

The main nuclear fuel is uranium-235, found in uranium oxide. This ore occurs in the old rocks of Canada, southern Africa, and Australia, but reserves are strictly limited. Low grade uranium ores occur in black shales in many parts of the world, however, and although they are expensive to mine, they could produce a great deal more fuel when required. The use of breeder reactors can further reduce the dependence on uranium, because they also create plutonium-239, a fuel for more nuclear energy. It is unfortunate that nuclear power is hazardous: as well as the unlikely danger of the accidental release of nuclear contaminants from a reactor, there is the problem of the disposal of radioactive wastes, some with long half-lives, which are created in the reactor. These wastes can be either diluted in the oceans, so producing a long term hazard for man, or buried underground, but this could cause contamination of water supplies. An associated danger was realized when liquid wastes were pumped deep underground in Colorado; the liquid reduced friction in stressed rocks and triggered earthquakes. This incident was quickly recognized by seismologists as a way of preventing the build-up of earth stresses, reducing the chance of violent earthquakes.

Nuclear power station *Inset* Relationship between disposal of liquid waste in a disposal well and earthquakes at a site in the Rocky Mountains.

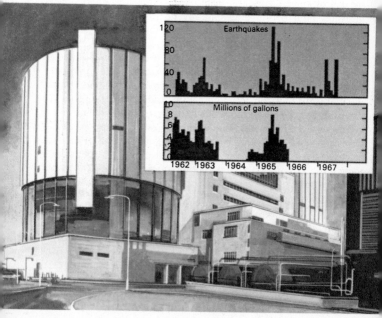

Economic minerals

Workable concentrations of minerals are called *ores*. The Earth's many elements normally lie scattered in the crust, but they have been concentrated in places by a variety of geological processes. Even ores of the same name have been produced in several ways; this is well illustrated by the ores of iron. The important magnetite (Fe_3O_4) iron ores of Sweden were produced by magmatic processes; the immense Ukrainian haematite (Fe_2O_3) and other iron ores are metamorphosed Precambrian sedimentary iron deposits; the Jurassic iron ores of Britain are marine sedimentary deposits and their composition ranges from hydrated oxides (limonite – $2Fe_2O_3.3H_2O$), carbonates (siderite – $FeCO_3$) to complex silicates such as chamosite; the siderite and iron sulphide (pyrite – FeS_2) ores of Bilbao in Spain are the result of mineral replacement in a limestone. Other processes have led to other iron concentrations elsewhere. The same variety applies to the ores of other minerals and some of these will be described briefly in the following pages.

Low grade iron ore in Western Australia.

Many of the world's principal non-ferrous ores are the result of igneous activity. Important concentrations of copper, lead, zinc, tin, mercury, and silver are deposited from hot aqueous (hydrothermal) solutions which emanated from cooling igneous bodies to give either *veins* or low quality disseminated ore bodies often of vast dimensions. Veins are thin discontinuous sheets of ore minerals and *gangue* minerals, the latter consisting chiefly of quartz and carbonate minerals of no economic value. The veins occupy hollows in fault, joint, and bedding planes where many of the minerals develop good crystal shapes. Particularly thick veins or groups of veins are *lodes*.

The disseminated ore bodies consist of a host rock in which the hydrothermal solutions precipitated ore minerals at widely scattered intervals. Such an ore body is said to be of low grade, because great quantities of rock must be worked to obtain the ore minerals. Vein ores are usually mined, the mine tunnels following the veins, while low grade ores are usually quarried, whole mountains being removed in the process.

Some important ore bodies are associated with the metamorphism

Above Section across a banded vein.
Below Concentration of ore deposits near a large intrusion.

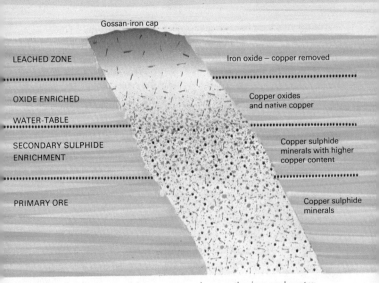

Gossan-iron cap

LEACHED ZONE   Iron oxide – copper removed

OXIDE ENRICHED   Copper oxides
                                                and native copper
WATER-TABLE

SECONDARY SULPHIDE   Copper sulphide
ENRICHMENT   minerals with higher
                                                copper content

PRIMARY ORE   Copper sulphide
                                                minerals

Enrichment of a low grade copper ore by weathering and water movements.

of sedimentary rocks, particularly limestones, by igneous intrusions. In such instances, gases have emanated from the igneous body depositing ores including those of copper, iron, zinc, and lead, in the newly created voids of the metamorphosed limestone. This process is correctly called *metasomatism* rather than metamorphism.

The settling of denser minerals within a cooling magma was described earlier in this book (page 16), and the same process can lead to the differentiation of minerals of economic value. The chromite ores of the Bushveld, South Africa, were formed this way, as were the concentrations of nickel sulphide of the Sudbury ore body in Ontario, Canada. In both cases, the accumulations are within basic igneous bodies.

Many ore bodies of low grade are enriched by solution and redeposition of original ore minerals by ground water. This natural up-grading can increase the value of a low grade ore body. The process is limited to the zone of weathering and ground water movement; with depth the value of the enriched ore will fall.

The principal non-ferrous ore fields of the British Isles – in Cornwall, northern England, Wales, and Ireland – are associated with exposed or inferred granite bodies which were the source of hydrothermal solutions.

Sedimentary processes give rise to accumulations of economic importance; such deposits include peat, coal and oil, several varieties of iron and manganese ore, some copper and phosphate deposits, clays, and salt deposits of various compositions. Many of these deposits, such as coal, are really only varieties of sedimentary rocks which happen to be of value to man. Oil is the result of organic sedimentation and subsequent migration to a porous, usually sedimentary, reservoir rock. Sedimentary phosphate deposits – a valuable source of fertilizer – accumulate on land or in marine conditions and are the result of the sedimentary concentration of organo-phosphates. These and many other substances of economic value found in sedimentary successions are the result of biochemical sedimentation when clastic sedimentation was absent or at a low rate, permitting accumulations of economic proportions.

A whole range of economically valuable rocks are the result of direct precipitation from sea-water by evaporation either in basins cut off from the sea, in shallow coastal zones, or in man-made salt 'pans'. Other evaporite deposits can occur when lakes dry out. Again, clastic sedimentation must be absent or minimal for valuable

Left Underground working of ancient salt deposits.
Right Present day salt pans.

Left Manganese nodules on ocean floor.
Right Structure of nodules.

5 cm

evaporite deposits to occur. Gypsum ($CaSO_4 \cdot nH_2O$) and anhydrite ($CaSO_4$) sulphates used in a variety of industries, are early precipitates from evaporating sea-water; with more intense evaporation, halite (common salt – NaCl) is precipitated. In special circumstances the body of evaporating sea-water can precipitate potash and other salts of even greater value to the chemical industry.

Vast accumulations of evaporites occur in the geological succession. These evaporites are either gallery mined or, in the case of the more soluble salts, they are recovered by solution mining when hot water is pumped into the deposit to return to the surface as a brine. Many salt domes have important accumulations of sulphur associated with them, which is another important raw material for chemical industries, and this too is recovered by the hot water method. Lake evaporites tend to be varied in composition because they are controlled by the composition of their source rocks; borax is one of the many dried lake evaporite minerals.

In recent years much interest has been shown in the manganese nodules which lie on the ocean's floor. Besides manganese, they contain copper, cobalt, and nickel, important raw materials, but so far none have been recovered on an economic scale.

Clam dredger recovering sand and gravel deposits.

Mechanical concentration by sedimentary processes can lead to economically valuable deposits of minerals which, in their source rocks, would have been uneconomic because of their scattered distribution. High energy environments such as occur in rivers and beaches are sites where sediments are sorted, permitting the development of placer deposits. Minerals suitable for such concentration are those resistant to mechanical and chemical breakdown while possessing a higher than average specific gravity. Diamonds, cassiterite (tin ore – SnO_2), titanium minerals such as rutile and ilmenite, and monazite which contains thorium and other rare elements, are some of the minerals recovered from present day and geologically recent beach placers. Gold is recovered from beach and river placers.

Sedimentary processes are also responsible for the accumulation of sand and gravel deposits, much in demand as concrete aggregates. Many such deposits in north-west Europe were laid down from the meltwaters of Ice Age glaciers on river terraces and flood plains and on the sea floor. Over 100 million tons of sand and gravel are used in Britain each year and about one tenth of this is recovered from the sea floor. Sand deposits, often from Mesozoic rocks, are of importance for water filtration plants, for mould making, and for the glass industry. The many different types of clays are also of value. The use to which they are put depends upon their composition; brickmaking, oil filtration, chemical absorption, the manufacture of cement, and as an industrial filler, are only a few of their uses. Another clay, china clay, used in the ceramic, paper, and plastics industries, is not a sedimentary clay. It is derived by the hydrothermal alteration and the natural weathering of granite. Hard metamorphic rocks such as hornfels, and sedimentary rocks, both in bulk and in a crushed state, are in constantly increasing demand by the construction industry for large structures such as dams, for roadstone and as a concrete aggregate. Natural stone is still used for buildings, particularly rocks like Portland and Cotswold stones which are easily cut freestones of pleasant appearance. Limestones are required in large quantities by the iron and cement industries, and siliceous sandstones are used for furnace linings.

Weathering, too, can lead to economically important accumulations; among the most important is the aluminium ore, bauxite, produced by the breakdown of aluminium silicate rocks under conditions of tropical weathering.

China clay pits.

Sea-water

Sea-water is salty, containing between 3·2 and 3·7 per cent of salts in solution, with common salt making up four-fifths of this amount. Many different elements, including uranium and gold, are contained in sea-water, so that it has been described as a low grade ore body. Some of its dissolved salts are precipitated by natural evaporation and were so precipitated in the geological past. A few salts are also extracted industrially; much of the world's magnesium and bromine are recovered from sea-water. Unfortunately, the recovery of many elements from sea-water remains uneconomic at today's prices; for instance, it costs four times the market price to recover uranium from sea-water. Sea-water is, however, in growing demand as a source of fresh water. This is recovered around the world by a variety of distillation processes. Freshwater obtained in this way is expensive, but it will be distilled increasingly in Middle East oil states where water is scarcer than oil. It remains to be seen if the saturated brines produced in these processes will be used for the recovery of sea-water's various elements.

Distillation plant to obtain fresh water from sea water.

Dam site in southern India – note the great size of the structure.

Geology and civil engineering

Geological information is critical for the design and construction of many civil engineering undertakings. This is well illustrated in the construction of a surface water-reservoir. There are four parts which require geological investigation: the catchment area of the reservoir's streams; the reservoir; the dam; and the spillway. Geologists can provide information about run-off, erosion, and sediment transport rates in the catchment. They can survey the geological setting of the reservoir, checking if the banks will landslide and if water will percolate out of the reservoir. The dam site requires special attention: Is there a good weight-bearing foundation rock? How deep is it? Can water percolate under the dam? Are there planes of weakness which will allow the dam and surrounding rocks to move when the reservoir is full? The geology of the site and the availability and nature of construction materials will influence the dam design. Careful geological and geophysical surveys aided by bore-hole data will provide many answers. Downstream, the river flow will be reduced, but near the dam the sediment-free, high velocity water will be capable of rapid erosion which could endanger

Folkstone Warren landslide, 1915.

the dam's foundations unless precautions, often of a geological nature, are undertaken.

The value of geological knowledge to many other engineering activities is obvious, but all too often there has been a lack of communication between engineer and geologist. In recent years a whole new field of engineering geology has developed which aims at making the best geological advice available. There are many instances where geological factors have proved to be critical. The Vaiont dam disaster in northern Italy in 1963 was caused by a large landslide in geologically unstable conditions. Alternating clays, marls, and limestones began to slide into the reservoir soon after its completion. Such was the size of the slide that every monitoring device failed to reach the slide plane. When the rate of movement became obvious, the water level was lowered, but this worsened the situation, because it left heavily waterlogged rocks supporting the slope. The ensuing rapid slide pushed the waters over the high dam which withstood the shock, but the floodwaters killed over 2000 people.

At best any slope can only be described as 'apparently stable', and regions once affected by glaciation are particularly prone to landsliding. Impeded drainage can cause slides, because water increases the weight of the slope and reduces the frictional resistance to sliding. Engineering works can lessen slope stability by excavation of the toe of a slope or by adding to its load.

Engineers can prevent landsliding by remedial work. At Folkestone Warren, Kent in England, where the main railway line runs near the coast, the chalk had often slid seawards on the underlying Gault clay as the sea eroded the toe of the slope. By rebuilding the sea wall, loading the front of the slide with concrete, and improving the drainage, engineers have effectively prevented further slides. At the same time they have installed a variety of warning devices in case renewed landsliding should occur.

Tunnels are another aspect of engineering where geological knowledge is important. The geologist tries to inform the engineer of the rock types to be encountered during excavation, whether water flows can be expected, what stresses exist in the rocks, how the newly exposed rocks will react to the atmosphere, and many other factors. All the methods of geological and geophysical investigation must be used to best advantage, because tunnels are usually required in places where underground evidence is hard to obtain such as under mountains, towns, rivers, and, as in the case of the proposed Channel tunnel, under the sea.

The Channel Tunnel.

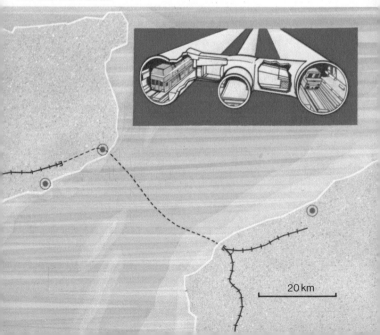

20 km

Geological surveys and societies

Nearly every country has a government agency charged with investigating that country's geology. The first such government geological survey was founded in Great Britain in 1835 on the recommendation of Sir Charles Lyell. Britain's Geological Survey, or Institute of Geological Sciences, as it has been called since 1965, is concerned with making geological maps, investigating the subsurface geology of Britain and the adjacent continental shelf, storing all types of geological information and making such knowledge available through its information services, publications, and museum. It is also concerned with geological investigations overseas, often as aid to developing countries. Most of Britain has been mapped on a scale of six inches to one mile, and these original maps may be consulted. Geological maps on a scale of one inch to one mile and more recently on a scale of 1 : 50 000 may be purchased, as may smaller scale maps covering larger areas. A particularly important and attractive part of the Institute's activities is the museum where all aspects of geology are on display.

Much original geological research is performed by teachers and students of geology who work in universities. Almost every university has a department of geology in its science faculty, and each year several hundred students graduate in geology in Britain.

Part of the museum at the Institute of Geological Sciences, London.

The Geological Society, Piccadilly, London.

Many of these graduates find employment in oil and mining companies, while others take work in civil engineering companies or government agencies. Most graduate geologists have the opportunity to work overseas; indeed, travel and adventure attract many young people to the subject. Today there are many opportunities for female geologists and many have achieved distinction for their work.

Geology, like many other subjects, has its learned societies. The Geological Society in London, founded in 1807, was the first society for geologists. This society, and others like it, provides a meeting place for its Fellows, has a lecture theatre, a large library of geological books and maps, and publishes journals, memoirs, and reports. Many countries have a national geological society, but there are many societies, associations, and clubs which are concerned with the more local aspects of geology and which publish useful geological guides and arrange excursions.

CONCLUSION

This book cannot do more than introduce you to geology. It has attempted to describe something of the great breadth of the subject, but it would be unfortunate if you felt that the subject is too remote or too enormous in which to get involved. One of the main attractions of geology is that it can be an exciting and worthwhile hobby. A knowledge of geology can be used to appreciate a little more fully the natural beauty of the surrounding countryside – scenery is very much the product of geological processes – and to gain pleasure in finding minerals, rocks, and fossils at a rock exposure or on the beach. It may be that observing or collecting is satisfaction enough, but remember that amateur geologists have consistently made worthy scientific discoveries in this subject and will no doubt continue to do so.

The illustration shows some of the equipment used by a professional geologist, but it is not necessary to be so equipped to pursue some gentle geology; indeed, one item that is often overworked by the would-be geologist is the hammer! Start with a visit to the local library to seek some background information; if there

Some items of equipment used by the field geologist.

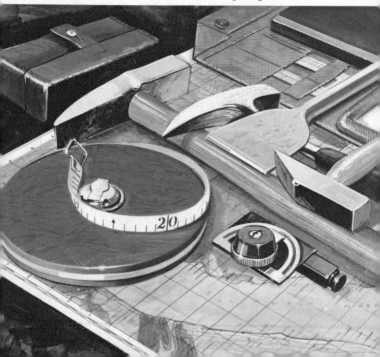

is a municipal museum pay it several visits to learn something of the regional geology. Arrange to visit one or two rock exposures, if possible with somebody who has some geological experience. Always remember that many exposures are on private land and that permission must be sought before going on to it: learn the Country Code and abide by it. Old and working quarries provide interesting exposures, but these are often dangerous places, and casual visitors are not usually welcome in working quarries. When you have reached an exposure do not indulge in indiscriminate hammering and chiselling of the rock – many of the best fossils and minerals lie on waste tips and in weathered debris. If hammering is necessary, do not leave fragments where they can cause injury to other persons or to animals.

A great deal of interesting geology lies in the wilder parts of any country, in areas of mountains or high cliffs. Care must be taken to be well equipped before visiting such places: avoid going alone, or make sure others know where you plan to go, because absorbing geology can often make an inexperienced person unaware of either the lowering cloud or the rising tide.

BOOKS TO READ

Elementary books with abundant colour and monochrome illustrations.

Bertin, Leon. 1972. *The New Larousse Encyclopedia of the Earth.* London, Hamlyn. A comprehensive treatment of all aspects of geology with hundreds of illustrations.

Calder, Nigel. 1972. *The Restless Earth.* London, BBC Publications Ltd. Earth movements and plate tectonics described with many illustrations.

Evans, I. O. 1970. *The Earth.* London, Hamlyn. A readable account of the history of geology.

Guest, J. (Editor). 1971. *Earth and its Satellite.* London, Hart-Davis. A lavishly illustrated symposium on the Earth and the Moon.

Harris, Reg. 1969. *Natural History Collecting.* London, Hamlyn. A handy book on collecting methods, with a useful section on collecting rocks, minerals, and fossils.

Institute of Geographical Sciences. 1972. *The Story of the Earth.* Her Majesty's Stationary Office. An inexpensive, authoritative booklet prepared in conjunction with the recently opened exhibition at the Institute of Geological Sciences.

Zim, H. S. and Shaffer, P. R. 1965. *Rocks and Minerals.* London, Paul Hamlyn. An inexpensive general guide to the main minerals and rocks.

Zim, H. S. and Shaffer, P. R. 1965. *Fossils.* London, Paul Hamlyn. A companion work to *Rocks and Minerals* which introduces fossils.

Further reading: more advanced books.

Gass, I. G., Smith, P. T. and Wilson, R. C. L. 1971. *Understanding the Earth.* Open University Press. Prepared as the basic book for the Open University Foundation Course in Geology. Contains papers on most recent developments.

Holmes, A. 1965. *Principles of Physical Geology.* London, Nelson. The last edition of what is perhaps the most authoritative and readable book on physical geology.

Scientific American, 1971. *Energy and Power.* San Francisco, W. H. Freeman & Co. Series of articles on energy developments and requirements.

Whitten, D. G. A. and Brooks, J. R. V. 1972. *The Penguin Dictionary of Geology.* Harmondsworth, Penguin Reference Series.

Wilson, J. T. (Introduction) Scientific American. 1972. *The Continents Adrift.* San Francisco, W. H. Freeman & Co. A series of papers which illustrate the development of ideas on continental drift and plate tectonics.

INDEX

SOME OTHER TITLES IN THIS SERIES

Arts
Antique Furniture/Architecture/Art Nouveau for Collectors/Clocks and Watches/Glass for Collectors/Jewellery/Musical Instruments/Porcelain/Pottery/Silver for Collectors/Victoriana

Domestic Animals and Pets
Budgerigars/Cats/Dog Care/Dogs/Horses and Ponies/Pet Birds/Pets for Children/Tropical Freshwater Aquaria/Tropical Marine Aquaria

Domestic Science
Flower Arranging

Gardening
Chrysanthemums/Garden Flowers/Garden Shrubs/House Plants/Plants for Small Gardens/Roses

General Information
Aircraft/Arms and Armour/Coins and Medals/Espionage/Flags/Fortune Telling/Freshwater Fishing/Guns/Military Uniforms/Motor Boats and Boating/National Costumes of the world/Orders and Decorations/Rockets and Missiles/Sailing/Sailing Ships and Sailing Craft/Sea Fishing/Trains/Veteran and Vintage Cars/Warships

History and Mythology
Age of Shakespeare/Archaeology/Discovery of: Africa/The American West/Australia/Japan/North America/South America/Great Land Battles/Great Naval Battles/Myths and Legends of: Africa/Ancient Egypt/Ancient Greece/Ancient Rome/India/The South Seas/Witchcraft and Black Magic

Natural History
The Animal Kingdom/Animals of Australia and New Zealand/Animals of Southern Asia/Bird Behaviour/Birds of Prey/Butterflies/Evolution of Life/Fishes of the world/Fossil Man/A Guide to the Seashore/Life in the Sea/Mammals of the world/Monkeys and Apes/Natural History Collecting/The Plant Kingdom/Prehistoric Animals/Seabirds/Seashells/Snakes of the world/Trees of the world/Tropical Birds/Wild Cats

Popular Science
Astronomy/Atomic Energy/Chemistry/Computers at Work/The Earth/Electricity/Electronics/Exploring the Planets/Heredity/The Human Body/Mathematics/Microscopes and Microscopic Life/Physics/Psychology/Undersea Exploration/The Weather Guide